Oxford Cambridge and RSA Examinations

OCR
RECOGNISING ACHIEVEMENT

# Entry Level Mathematics

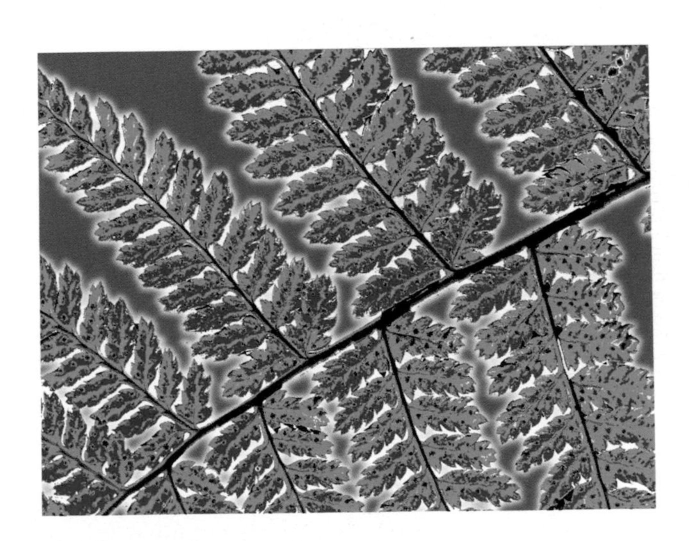

SERIES EDITOR BRIAN SEAGER

CHRISTINE WATSON, HEATHER WEST

Hodder & Stoughton
A MEMBER OF THE HODDER HEADLINE GROUP

Every effort has been made to trace ownership of copyright. The Publishers would be happy to make arrangements with any copyright holders whom it has not been possible to trace.

Orders: please contact Bookpoint Ltd, 130 Milton Park, Abingdon, Oxon OX14 4SB. Telephone: (44) 01235 827720. Fax: (44) 01235 400454. Lines are open from 9.00–6.00, Monday to Saturday, with a 24 hour message answering service.

*British Library Cataloguing in Publication Data*
A catalogue record for this title is available from the British Library

ISBN 0 340 801638

First published 2001
Impression number 10 9 8 7 6 5 4 3 2
2007 2007 2005 2004 2003 2002



Typeset by Hart McLeod, Cambridge
Illustrations by Phil Treble and Sarah Warburton

# Introduction

This student book and the accompanying resources cover levels 1 to 3 of the National Curriculum and are suitable for use with students working towards an Entry Level Certificate in Mathematics. It provides complete coverage of each OCR specification.

## Units

- The content is arranged into a Preliminary Unit and 20 further Units.
- Each unit contains topics at approximately the same level, covering a variety of mathematical areas.
- Topics are revisited and extended in later units.
- Where appropriate, topics are clearly identified as calculator or non-calculator.
- Students should start at the unit most appropriate to their level. Topics from earlier units can be used to support revision.

## Activities

- **Challenges** are mini-investigations and are included to extend understanding, and can be used as classwork or homework. It is not essential that all students complete them.
- **Worksheets** are an integral part of the content of each unit. Some involve practical activities, some allow easier access to the topic, some provide diagrams for students to work with.
- **Test Yourself** sheets are for use in pairs and encourage verbalisation of mathematics and mental skills.
- **Sums Practice** sheets provide opportunities to revisit arithmetic skills without context.
- **Games** reinforce understanding and use of language. They can be used when completing the topic or as later revision, and are adaptable to individual or group work.
- **Investigations** are included as revision and to develop skills necessary for Practical assessments.
- The **Teacher's Guide** contains suggestions for further practical activities, games and use of IT. Masters are included for Test Yourself and each Game to allow additional resources to be developed.

## Assessment

- **I can** sheets are diagnostic or end-of-unit assessments, ideal for inclusion in a Record of Achievement.

# Contents

## Preliminary Unit



## Unit 1



## Unit 2



## Unit 3



## Unit 4



## Unit 5



## Unit 6



## Unit 7



## Unit 8



## Unit 9

## Unit 10



## Unit 11



## Unit 12



## Unit 13



## Unit 14



## Unit 15



## Unit 16



## Unit 17



## Unit 18



## Unit 19



## Unit 20



## Reference

# Preliminary Unit

## Counting 1

### Counting to ten

▶ Copy and complete the missing numbers.

You can use the number list to help you.

| | | | | |
|---|---|---|---|---|
| **A** | 1 | 2 | 3 | ___ |
| **B** | 6 | 7 | 8 | ___ |
| **C** | 3 | 4 | ___ | 6 |
| **D** | 7 | 8 | ___ | 10 |
| **E** | 2 | ___ | 4 | 5 |
| **F** | five | four | three | ___ |
| **G** | four | ___ | six | seven |
| **H** | ten | nine | eight | ___ |
| **I** | two | ___ | four | ___ |
| **J** | ___ | three | two | ___ |

1 one ✦

2 two ✦✦

3 three ✦✦✦

4 four ✦✦ ✦✦

5 five ✦✦✦ ✦✦

6 six ✦✦✦ ✦✦✦

7 seven ✦✦✦ ✦ ✦✦✦

8 eight ✦✦✦ ✦✦ ✦✦✦

9 nine ✦✦✦ ✦✦✦ ✦✦✦

10 ten ✦✦✦ ✦✦ ✦✦✦ ✦✦

## Counting objects

There are **nine** boxes altogether.

You need to practise counting to 10 . . . or use worksheet P/1.

► How many are there in each box?

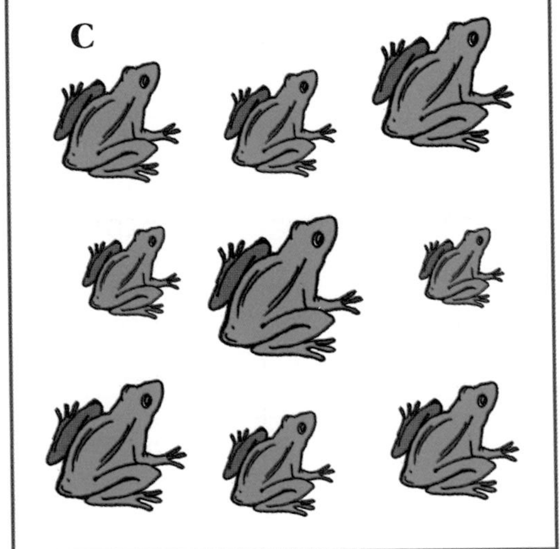

# Days and months

## Days of the week

These are the days of the week. You need to know the spelling and the short way to write them.

▶ Copy this table into your book.
*(. . . or use the Snap Cards game)*

| | |
|---|---|
| Monday | Mon |
| Tuesday | Tue |
| Wednesday | Wed |
| Thursday | Thur |
| Friday | Fri |
| Saturday | Sat |
| Sunday | Sun |

## Months of the year

These are the months of the year. You need to know the spelling and the short way to write them.

▶ Copy this table into your book.
*(. . . or use the Snap Cards game)*

| | |
|---|---|
| January | Jan |
| February | Feb |
| March | Mar |
| April | Apr |
| May | May |
| June | Jun |

| | |
|---|---|
| July | Jul |
| August | Aug |
| September | Sept |
| October | Oct |
| November | Nov |
| December | Dec |

## Before and after

▶ Use the tables to help you answer these questions.

Check your spellings.

**1** Which day comes **after Tuesday**?

**2** Which day comes after Fri?

**3** Which day comes before Thursday?

**4** Which day comes after Sun?

**5** Which day comes before Wed?

**6** There are two days at the weekend.
Write their names the short way.

**7** Which month comes after March?

**8** Which month comes after Oct?

**9** Which month comes before Feb?

**10** Which month comes after December?

**11** Which month is number 4?

**12** Which month comes before Sept?

**13** Which month comes before July?

**14** Which month is number 11?

**15** How many months are there in one year?

You need to learn the shorthand for days and months . . .
or use worksheet P/2.

# Shapes 1

## Straight or curved?

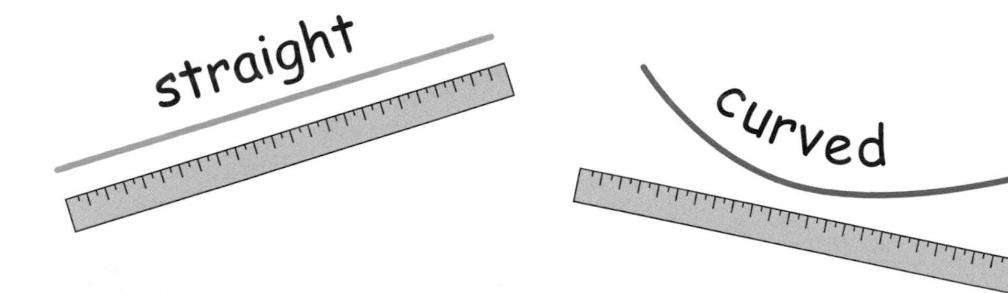


► Are these curved or straight?
Check with a ruler if you are not sure.

For each one, write down curved or straight.

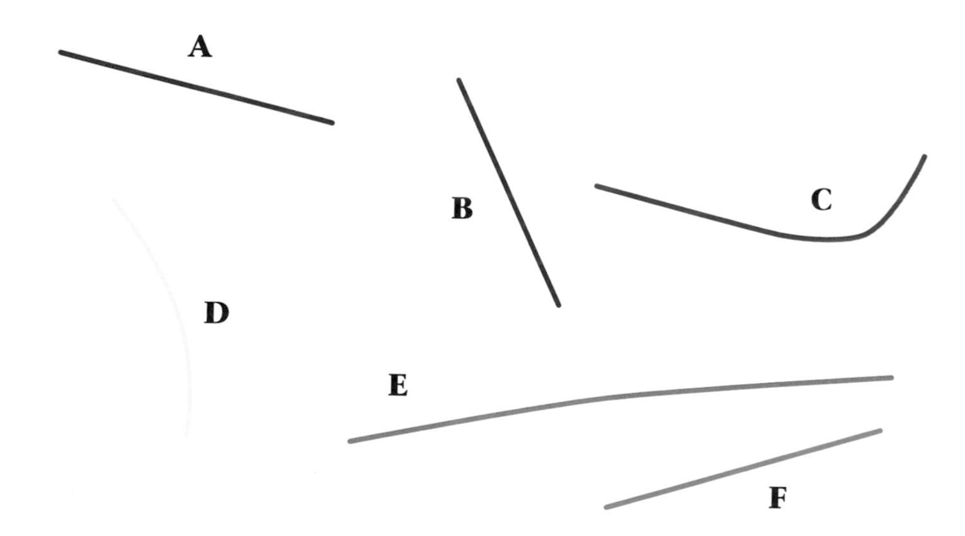


## Sides and corners

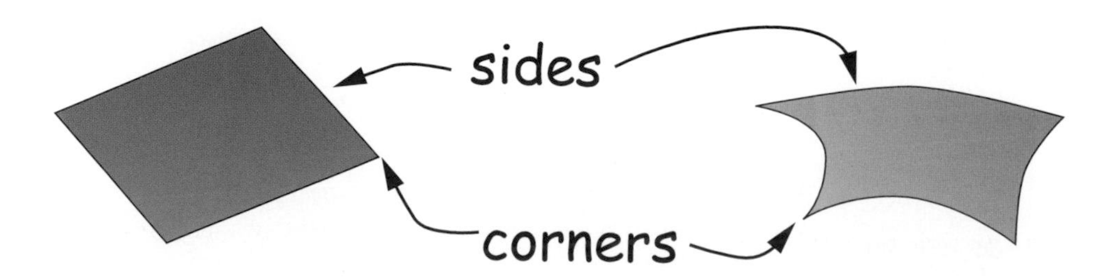

▶ Copy these shapes and complete the sentences.

It has __3__ straight sides.

It has ______ corners.

It has ______ curved sides.

It has ______ straight sides.

It has ______ corners.

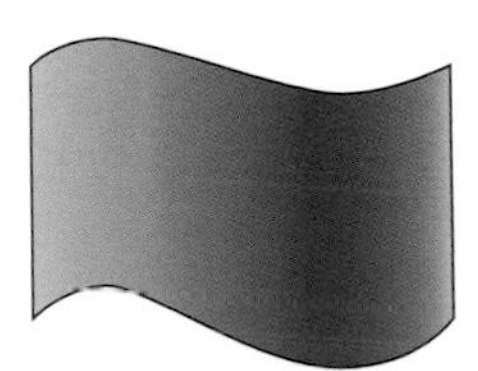

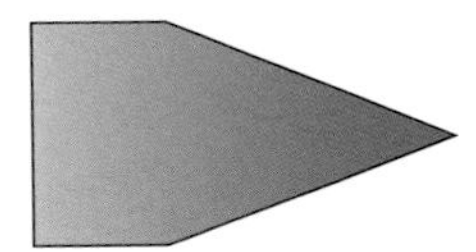

It has ______ straight sides.

It has ______ corners.

It has ______ curved sides.

It has ______ corners.

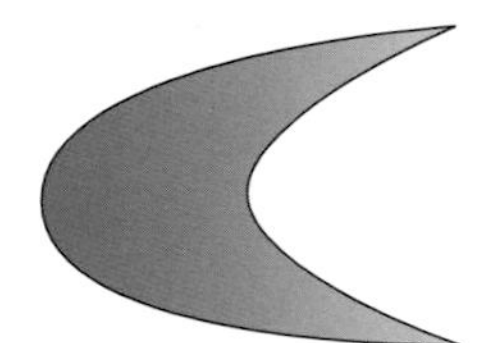

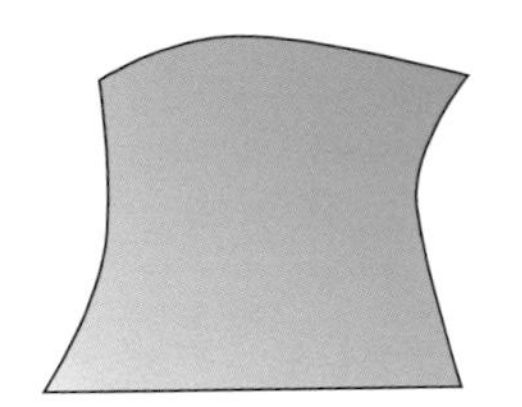

It has ______ curved sides.

It has ______ straight sides.

It has ______ corners.

It has ______ curved sides.

It has ______ straight sides.

It has ______ corners.

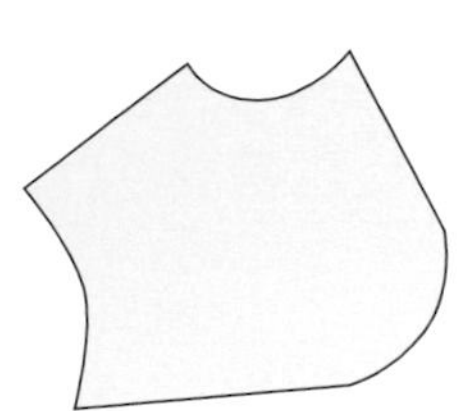

 You need to practise sorting shapes . . . or use worksheet P/3.

# Adding 1

*you may use cubes or counters*

**4 + 5 = 9**

 + = 

► Use counters to help you to do these sums.

| | | | | | |
|---|---|---|---|---|---|
| **A** | 1 + 3 | **H** | 1 + 9 | **O** | 5 + 5 |
| **B** | 2 + 2 | **I** | 2 + 5 | **P** | 2 + 6 |
| **C** | 4 + 1 | **J** | 4 + 6 | **Q** | 7 + 3 |
| **D** | 3 + 2 | **K** | 5 + 3 | **R** | 5 + 4 |
| **E** | 7 + 1 | **L** | 3 + 4 | **S** | 8 + 2 |
| **F** | 1 + 5 | **M** | 2 + 7 | **T** | 3 + 6 |
| **G** | 4 + 4 | **N** | 6 + 4 | | |

## Challenge

\+ means **add**

How many other words can you find that mean + ?

# Clocks 1

The long hand on this clock is pointing to the 12.

The time is [ ] o'clock.

The short hand on this clock is pointing to the 10.

The time is [ 10 ] o'clock.

▶ What time does each clock show?

**A**

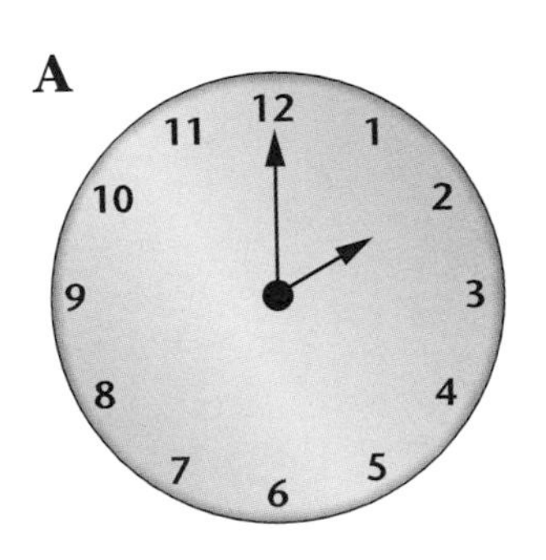


**B**

**C**

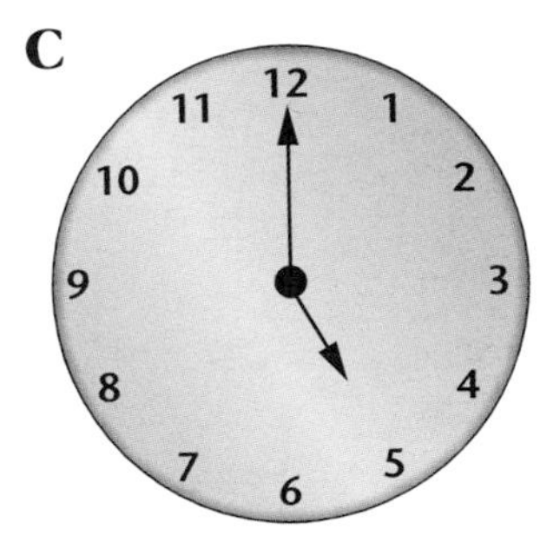


**D**

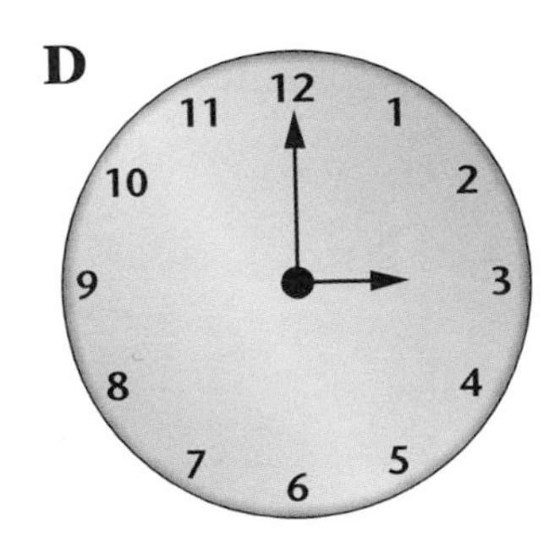


**E**

**F**

**G**

**H**

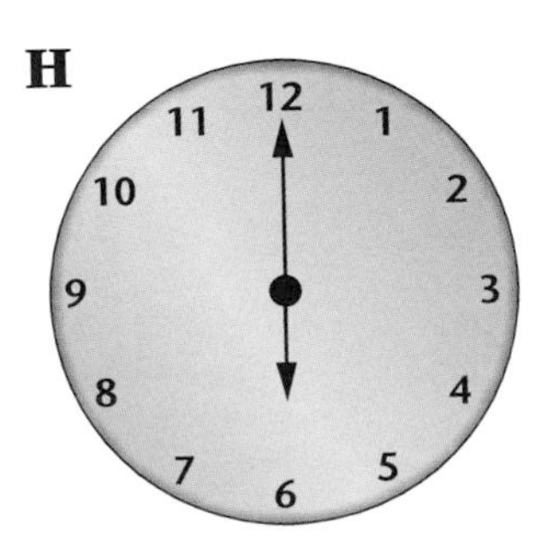


**I**

# Sorting 1

▶ These people have been sorted into two groups: **people standing** and **people sitting**.

**1** Which group should each of these people go into?

▶ These flags have been sorted into three groups: flags **with crosses**, flags **with stripes** and flags **with neither** crosses nor stripes.

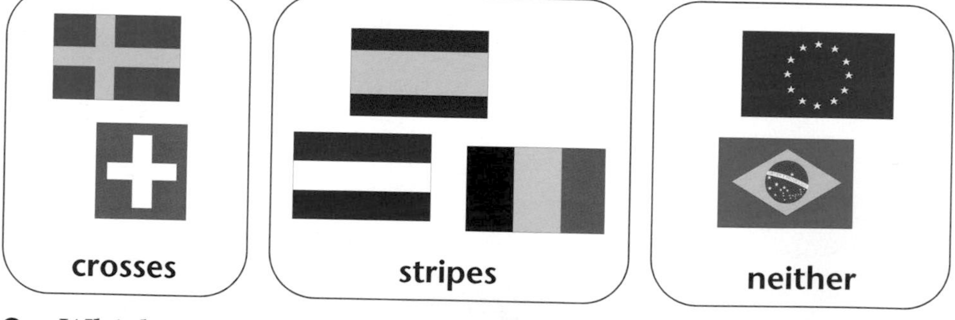


**2** Which group should each of these flags go into?

You need to practise sorting objects into groups . . . or use worksheet P/4.

# Counting 2

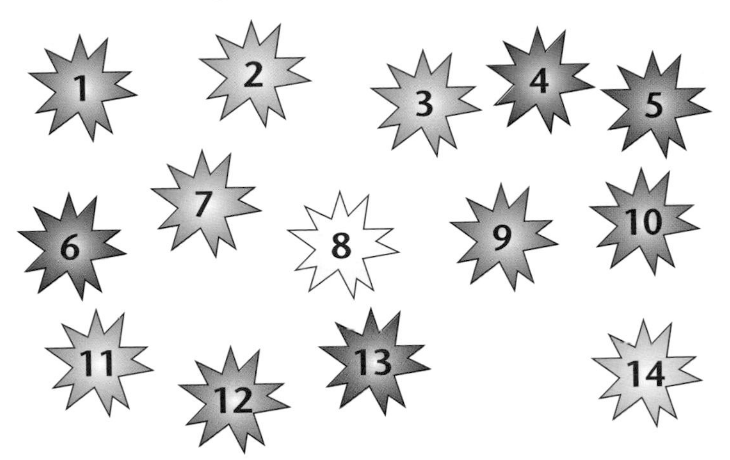


There are **fourteen** stars altogether.

You need to practise counting to 20 . . . or use worksheet P/5.

► How many are there in each box?

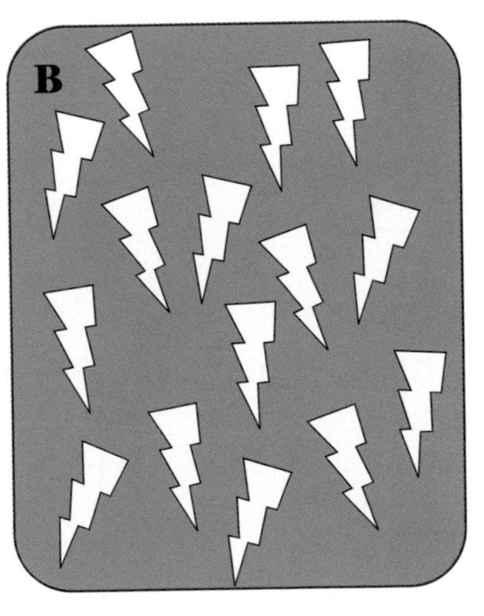

0 zero
1 one
2 two
3 three
4 four
5 five
6 six
7 seven
8 eight
9 nine
10 ten
11 eleven
12 twelve
13 thirteen
14 fourteen
15 fifteen
16 sixteen
17 seventeen
18 eighteen
19 nineteen
20 twenty

# Drawing shapes 1

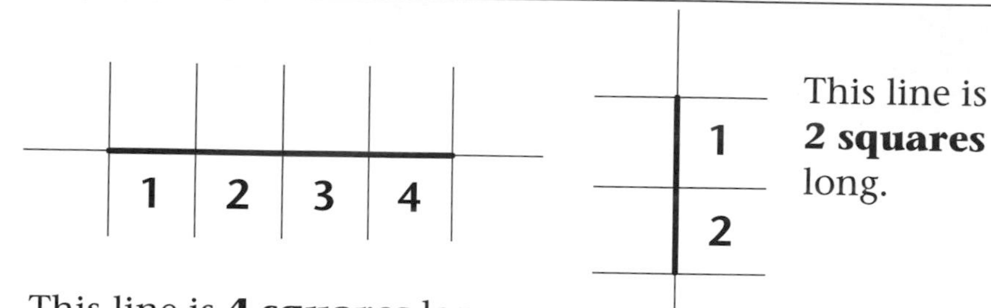


This line is **2 squares** long.

This line is **4 squares** long.

► Draw these lines on squared paper.
Make sure that your line is on top of one of the grid lines.

**A** 3 squares **B** 5 squares **C** 1 square **D** 8 squares

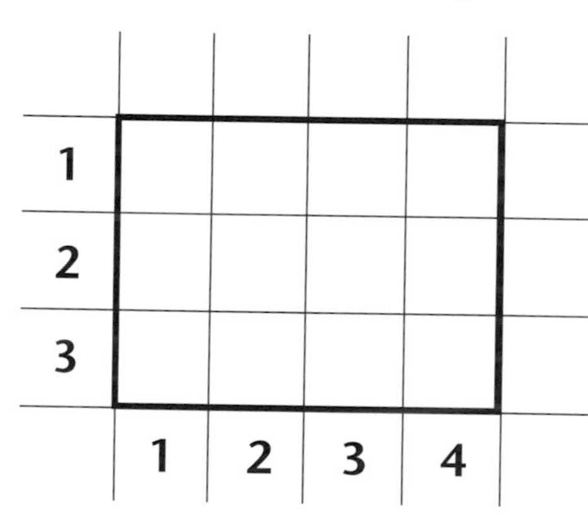


This shape is **4 squares** wide and **3 squares** tall.

► Copy this shape carefully onto squared paper.

► Copy this shape onto squared paper.

► Copy and complete this sentence.

This shape is

___ **squares** wide

and ___ **squares** tall.

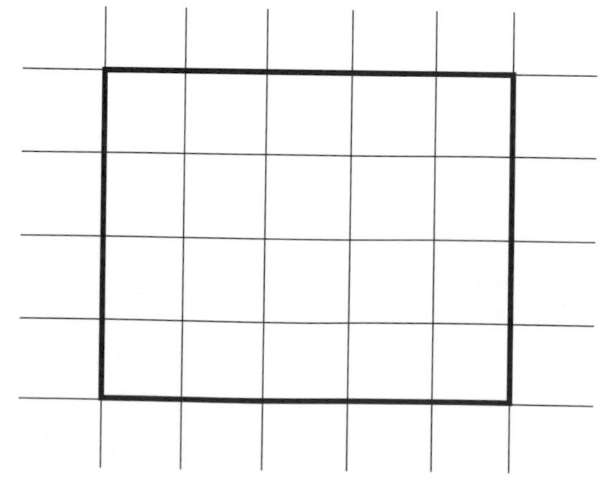

 You need to practise copying shapes onto squared paper . . . or use worksheet P/6.

# Unit 1

## Subtracting 1

**5 − 2 = 3**

 − = 

► Use counters to help you to do these sums.

**A** 3 − 1

**B** 5 − 4

**C** 4 − 2

**D** 6 − 1

**E** 5 − 3

**F** 9 − 1

**G** 7 − 2

**H** 3 − 2

**I** 6 − 5

**J** 8 − 3

**K** 6 − 2

**L** 10 − 5

**M** 7 − 6

**N** 9 − 4

**O** 6 − 4

**P** 8 − 5

**Q** 7 − 3

**R** 10 − 4

**S** 8 − 6

**T** 9 − 5

### Challenge

− means **subtract**

How many other words can you find that mean − ?

# Money 1

These are the tails of British coins.

► These are the heads of the coins. Write down the name of each coin.

▶ Write down how much money is in each of these pictures.

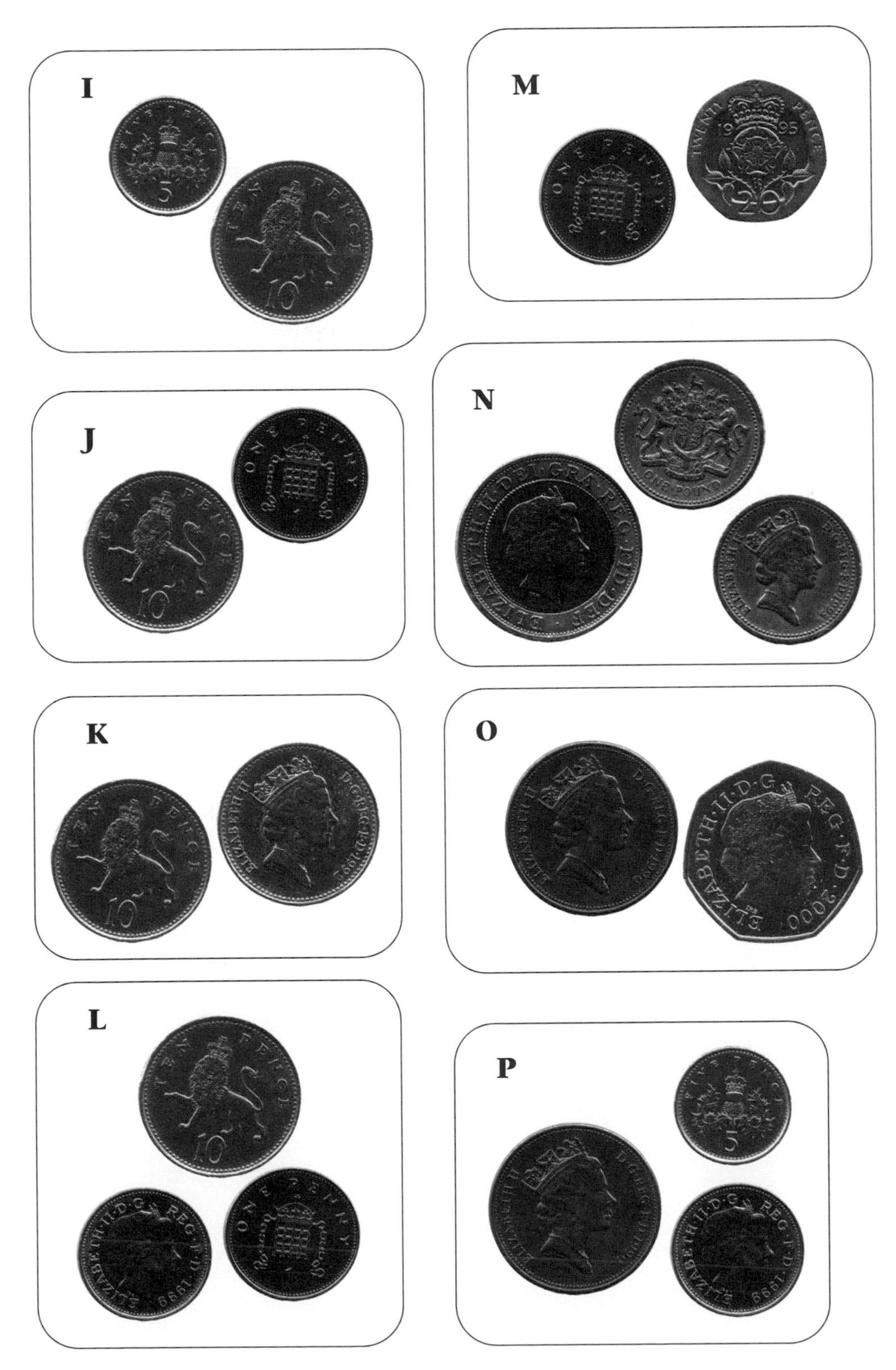

# Reading scales 1

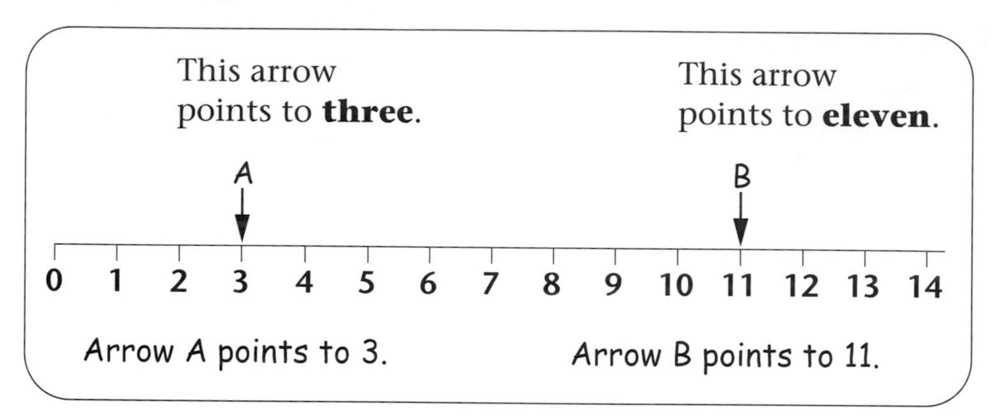


▶ Write down the number each arrow points to.

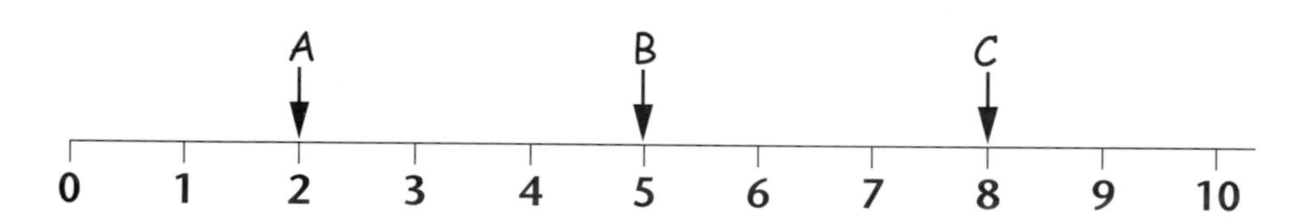


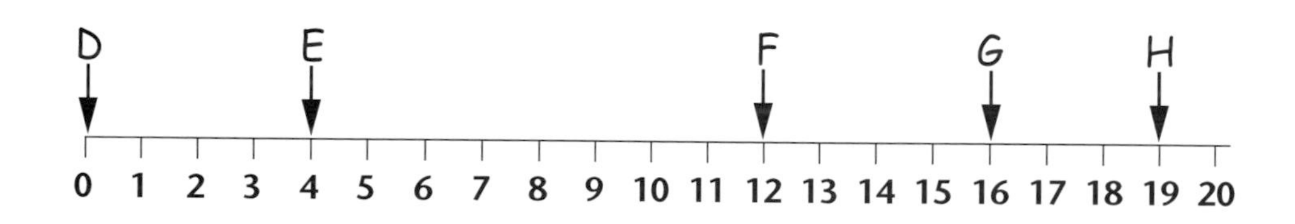


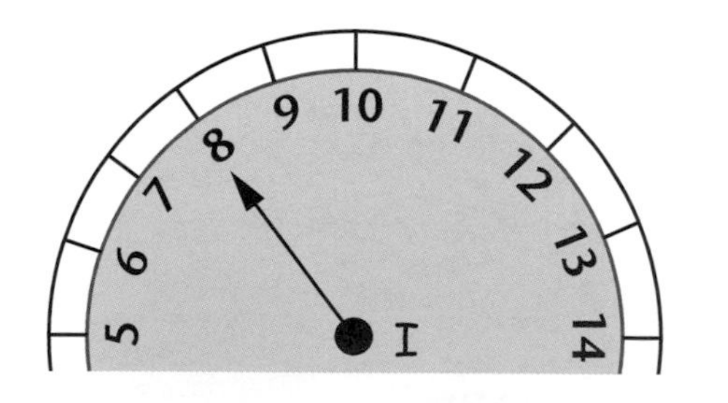


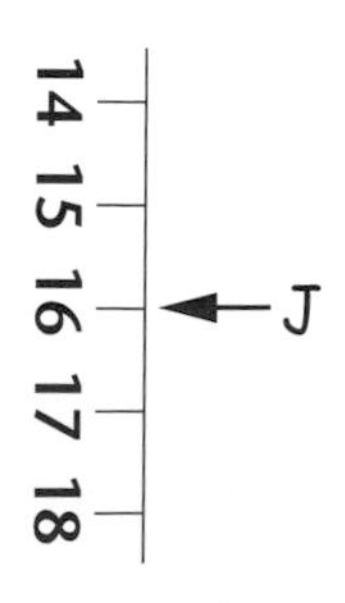

 You need to practise marking arrows on scales . . . or use worksheet 1/1.

# Missing numbers 1

▶ Copy and complete these sums.

There is a number missing from each □.

The first one is done for you.

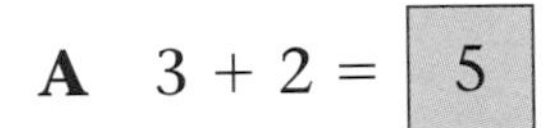

| | | | |
|---|---|---|---|
| **A** | 3 + 2 = 5 | 5 − 3 = 2 | 5 − 2 = 3 |
| **B** | 2 + 4 = □ | 6 − 4 = □ | 6 − 2 = □ |
| **C** | 1 + 7 = □ | □ − 7 = 1 | □ − 1 = 7 |
| **D** | 5 + 3 = □ | 8 − 3 = □ | 8 − □ = 3 |
| **E** | 2 + □ = 3 | 3 − □ = 1 | □ − 2 = 1 |
| **F** | □ + 6 = 7 | □ − 6 = 1 | 7 − □ = 6 |

▶ Each shape in these sums stands for a different number.
Copy and complete the sums.

| | | | |
|---|---|---|---|
| **G** | 6 + ✱ = 9 | 9 − ✱ = 6 | 9 − 6 = ✱ |
| **H** | ■ + 2 = 7 | ▼ − 2 = 5 | ▼ − 5 = 2 |
| **I** | 8 + ◆ = 10 | 10 − ◆ = 8 | ✦ − 8 = 2 |
| **J** | 4 + 6 = ✦ | ✦ − 6 = 4 | ✦ − 4 = 6 |
| **K** | ● + 7 = 9 | ▲ − 7 = 2 | 9 − ★ = 7 |
| **L** | 4 + 3 = ◗ | ◗ − 3 = 4 | ◗ − 4 = 3 |

# Repeating patterns 1

This is a repeating pattern.

**blue red red blue red red blue red red blue red** . . .

The next word in the pattern is **red**.

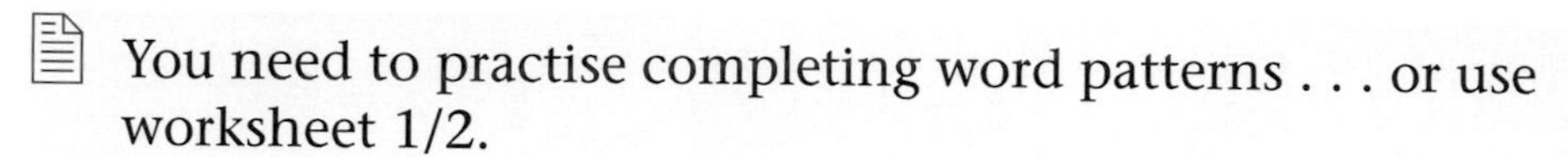

You need to practise completing word patterns . . . or use worksheet 1/2.

A repeating pattern can be put into groups.

(**blue red red**) (**blue red red**) (**blue red red**) (**blue red red**)

There are four groups in this pattern. It has **4 repeats**.

You need to practise counting the repeats in word patterns . . . or use worksheet 1/3.

## Challenge

*•••Emma••loves••Barry••Emma••loves••Barry•••*

▶ Design your own repeating necklace. It could be about a pop group or a friend.

# Add and subtract 1

*you may use the 100 line*

▶ Work these out without a calculator.

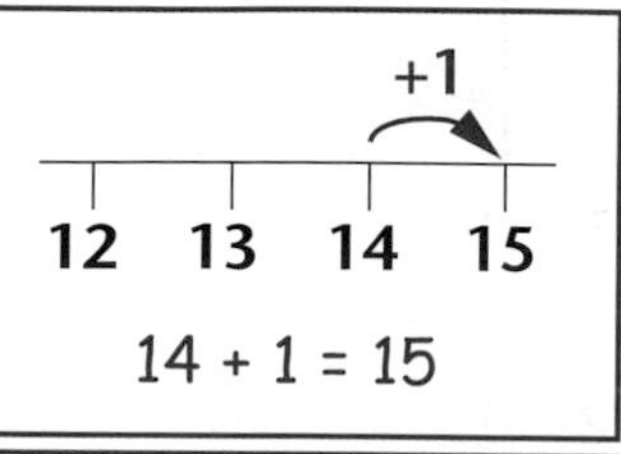


| | | | |
|---|---|---|---|
| **1** | 12 + 1 | **6** | 1 less than 6 |
| **2** | 16 − 1 | **7** | 1 more than 18 |
| **3** | 8 + 1 | **8** | 1 less than 17 |
| **4** | 19 − 1 | **9** | 1 more than 13 |
| **5** | 11 + 1 | **10** | 1 less than 20 |

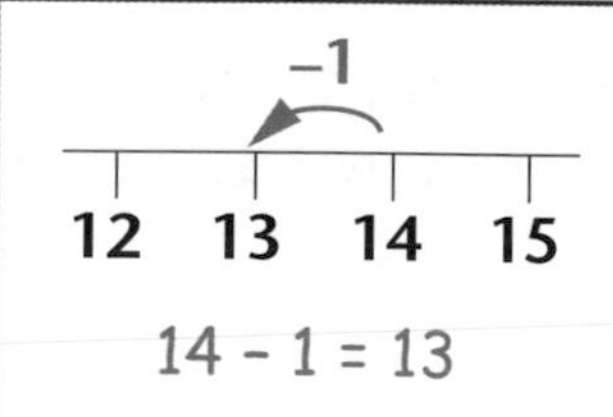

# Ted's tiles

*you need square dotty paper*

Ted has bought some tiles.

Each tile looks like this.

They are all the same size.

He needs to work out how to fit them together.

Ted decides to draw small blocks of pattern and choose the best one.

He starts with this grid of dots.

Ted draws these patterns to show ways of fitting the tiles together.

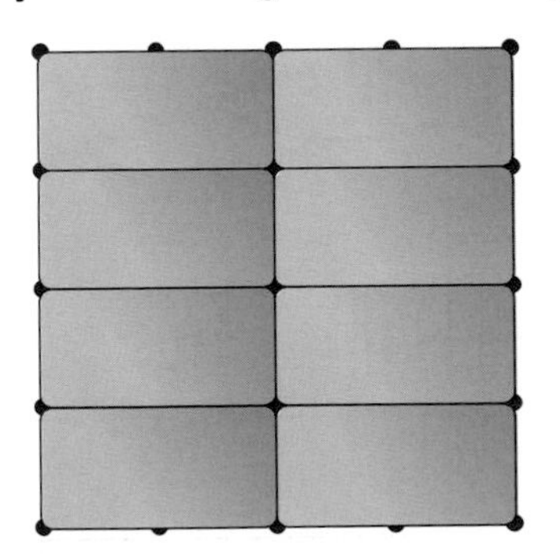

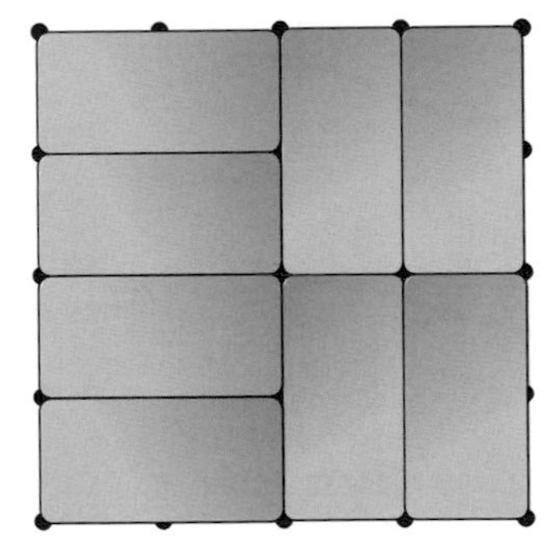

1 Draw some other ways of fitting eight tiles together in this block.

2 How many different ways can you find altogether?

3 Choose one of your patterns.
Fill part of a page with blocks of your pattern.

# Comparing lengths

*longer* *shorter* *taller* *longest* *shortest*

Scruffy and Rover have found sticks.

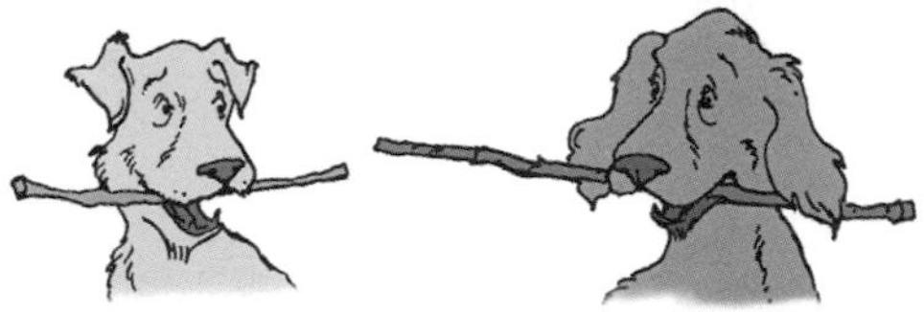

Scruffy Rover

Scruffy's stick
Rover's stick

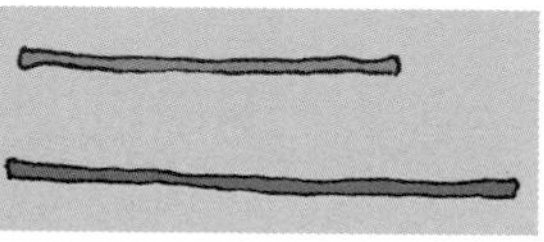

Scruffy's stick is **shorter**.
Rover's stick is **longer**.

▶ Which is **longer**?

1 A B

2 C D

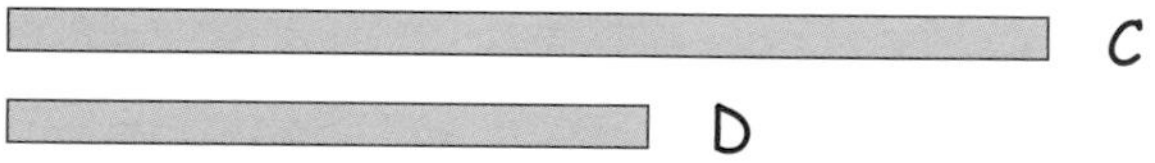

3 E F

4 G H

▶ Which is **shorter**?

5 J K

6 L M

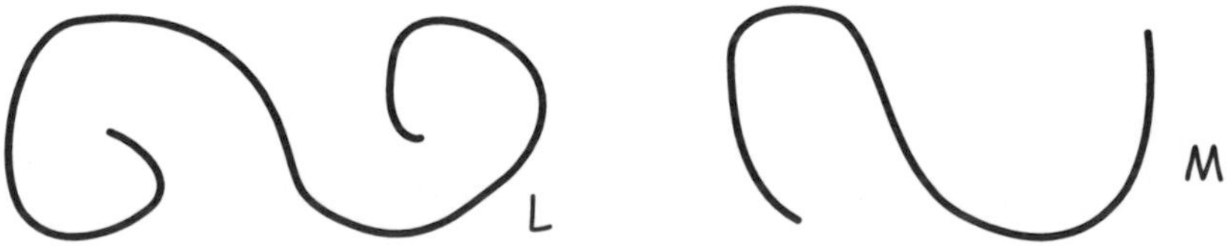

These children have picked straws to decide who goes first in a game.

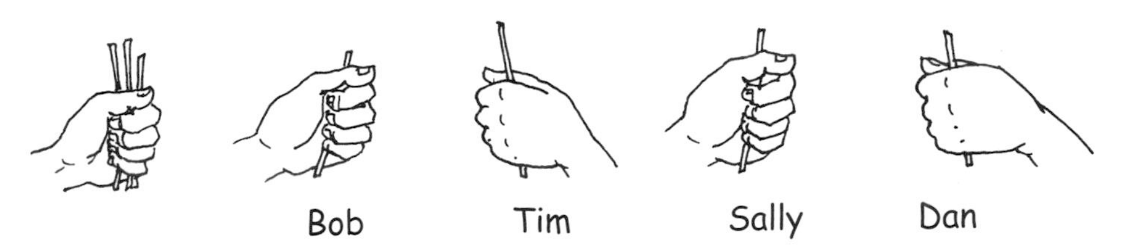


| | |
|---|---|
| Bob's straw | |
| Tim's straw | |
| Sally's straw | |
| Dan's straw | |

Sally has the **longest** straw.
She goes first.

Bob has drawn the **shortest** straw.
He goes last.

▶ Which is **longest**?

**7**

**8**

**9**

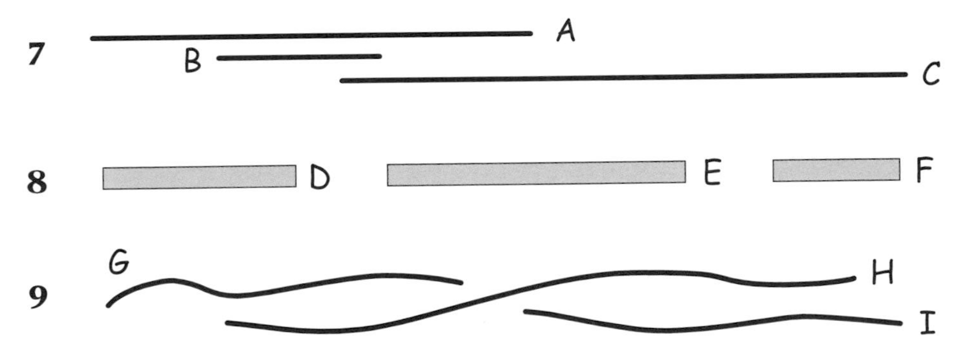


**10** **(a)** Which lines are **longer** than **N**?
**(b)** Which lines are **shorter** than **T**?

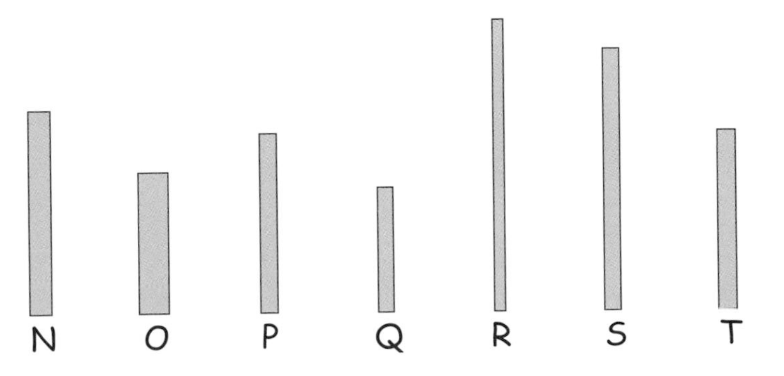

# Unit 2

## Shapes 2

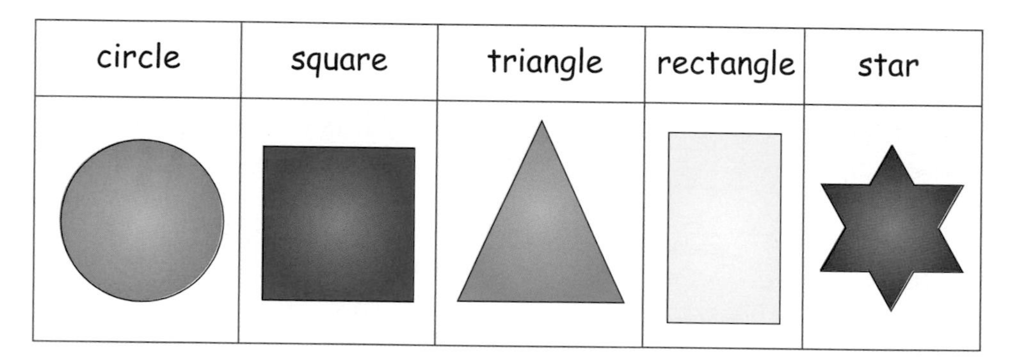


▶ Write down the name of each shape.

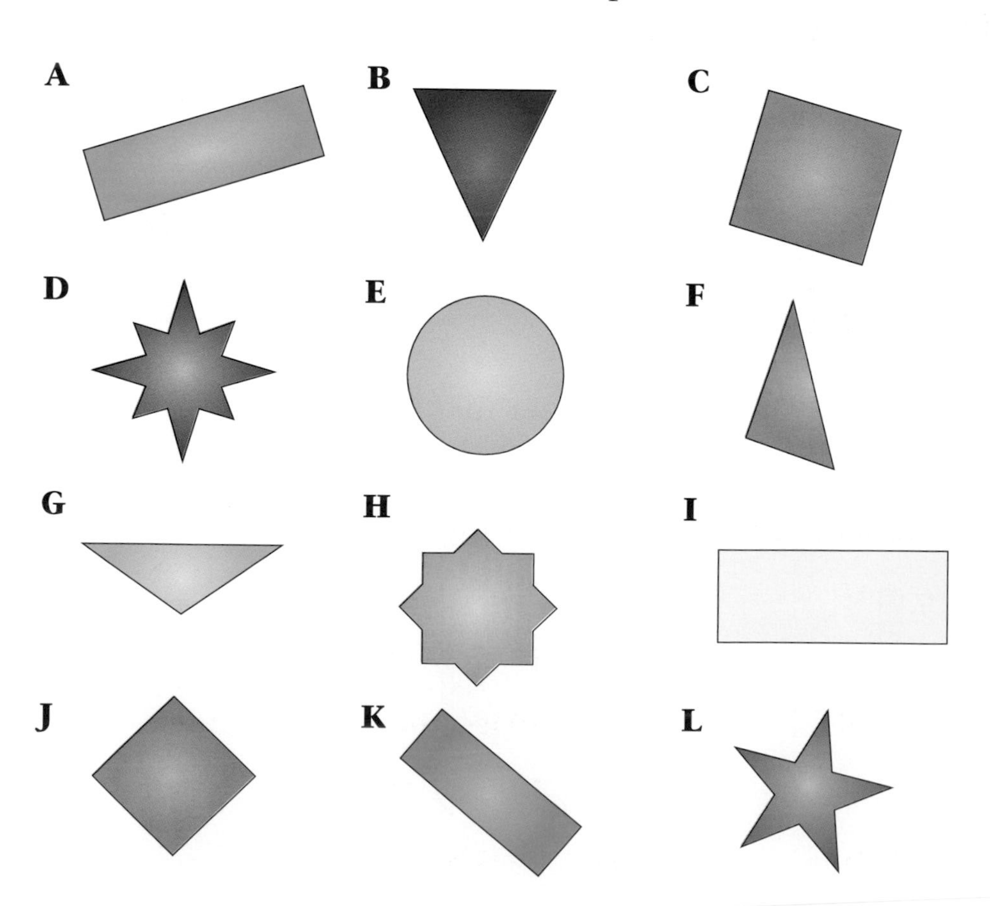

▶ Match up the ends of the sentences and copy them out.

| | |
|---|---|
| A circle is | 4 sides, equal length |
| A square has | 3 sides |
| A triangle has | 4 sides, opposite pairs equal length |
| A rectangle has | round |

## Challenge

Make a poster showing real objects which match these shapes.

circle ↔ coin    rectangle ↔ desk top

# Add and subtract 2

*you may use the 100 line*

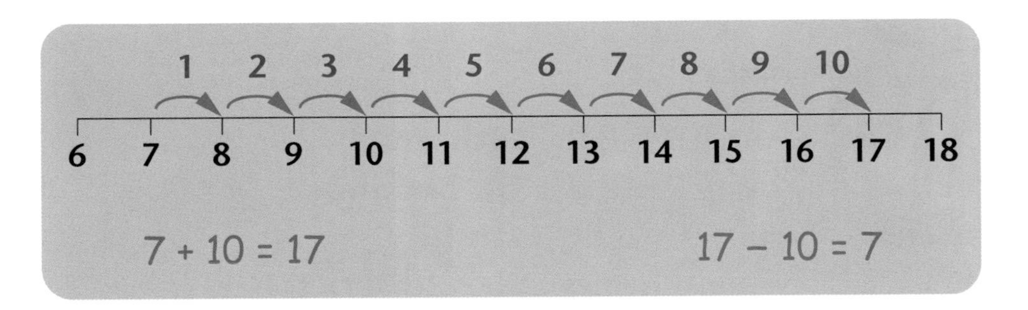


▶ Work these out without a calculator.

**1** 6 + 10

**2** 18 − 10

**3** 9 + 10

**4** 14 − 10

**5** 5 + 10

**6** 10 less than 12

**7** 10 more than 1

**8** 10 less than 13

**9** 10 more than 10

**10** 10 less than 19

# Dates

## Months

January is the **first** month in the year. It is month number **1**.
February is the **second** month in the year. It is month number **2**.

**1** Write a list of all the months.
Put the number next to each month.

January 1
February 2
March ?

**2** Match these months with their numbers.
Use your list to help you.

August May November

## Holidays

In America, the **4**th of **July** is a holiday called Independence Day.

The date can be written using shorthand: day/month **4/7**

▶ Copy and complete these sentences using shorthand for the dates.
*(. . . or use worksheet 2/1)*

**A** In Australia, Anzac Day is the **25th April**.
**B** In Haiti, Discovery Day is the **5th December**.
**C** In France, Bastille Day is the **14th July**.
**D** In Jamaica, National Heroes' Day is **20th October**.
**E** In Bangladesh, **26th March** is Independence Day.

## Years

The year can be written in shorthand.
The first two numbers are missed off.

| | |
|---|---|
| 2000 ⇨ 00 | 1965 ⇨ 65 |
| 2078 ⇨ 78 | 1984 ⇨ 84 |
| 2014 ⇨ 14 | 1906 ⇨ 06 |
| 2007 ⇨ 07 | 1950 ⇨ 50 |

▶ Write the shorthand for each of these years.

| | | |
|---|---|---|
| 1957 ⇨ ______ | 2036 ⇨ ______ | 1993 ⇨ ______ |
| 2015 ⇨ ______ | 1904 ⇨ ______ | 2003 ⇨ ______ |

## Dates

The month and year shorthand are put together to make dates.

**day / month / year**

Nora was born 18th April 1908

day 18 / month 4 / year 08 ⇨ 18/4/08

Peter was born 12th December 1945

day 12 / month 12 / year 45 ⇨ 12/12/45

▶ Copy and complete these sentences using shorthand for the dates.
*(. . . or use worksheet 2/1)*

**A** The first man walked on the moon on the **21st July 1969**.
**B** There was a solar eclipse on the **11th August 1999**.
**C** Roger Bannister ran one mile in under four minutes on the **6th May 1954**.
**D** Nelson Mandela was freed after 27 years in prison on the **11th February 1990**.
**E** The coronation of Queen Elizabeth was on the **3rd June 1953**.
**F** The first ITV television broadcast was on the **13th February 1955**.
**G** An earthquake in San Francisco on **April 18th 1906** lasted three minutes.

## Challenge

Elvis Presley was born on the 8th January 1935.
Queen Elizabeth was born on April 21st 1926.

▶ Do you know these dates of birth?
Find out as many as you can.

your own birthday
your parents (or the adults you live with)
the oldest person in your family    your best friend
your favourite pop star . . . footballer . . . sports star

# Position 1

| | |
|---|---|
| 1st | first |
| 2nd | second |
| 3rd | third |
| 4th | fourth |
| 5th | fifth |

Anne is **first** in the queue.
John is **second** in the queue.

▶ Answer these questions about this queue.

**1** Who is third in the queue?

**2** Who is fourth in the queue?

**3** Who is fifth in the queue?

▶ These runners are finishing a race.
They have their numbers on their vests.

**4** Which runner is coming first?

**5** Which runner is coming second?

**6** Which runner is coming third?

▶ Here are some instructions for making a cake.
They are mixed up.

| 1st | 2nd | 3rd | 4th |
|---|---|---|---|
| Mix the ingredients | Take cake out of oven | Weigh out ingredients | Put cake in oven |

**7** Put the instructions in the right order.

▶ Here is the alphabet.

A B C D E F G H I J K L M N O P Q R S T U V W X Y Z

**8** Which letter is first?

**9** What is the 5th letter of the alphabet?

**10** What is the 10th letter of the alphabet?

**11** What is the 17th letter of the alphabet?

▶ These children are playing a game.
They start with the youngest and take turns in order of their age.

**12** Who goes first?

**13** Who goes second?

**14** Who goes third?

**15** Who goes last?

Emma Age 4 | Joe Age 8 | Ken Age 6 | Laura Age 5

# Adding 2

▶ Work out the answers to these sums.
You can use cubes or counters to help if you want to.

**A** $4 + 2$

**B** $3 + 5$

**C** $7 + 1$

**D** $2 + 8$

**E** $5 + 5$

**F** $1 + 6 + 2$

**G** $7 + 3$

**H** $4 + 5 + 1$

**I** $2 + 2 + 2$

**J** $6 + 4$

Write down the sums that had the answer 10.
Can you find any more adding sums with the answer 10?

Harry is working out **7 + 2 + 3 + 8 + 1**

He looks for numbers that make 10 first.

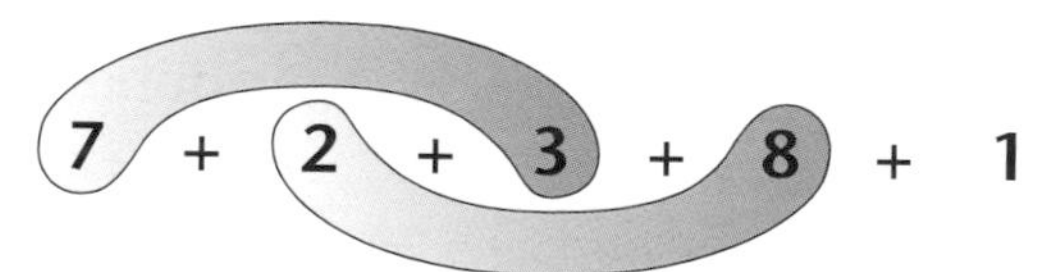


There are two tens and a one . . . the answer is $10 + 10 + 1 = 21$.

▶ Work out these sums. Look for tens.
Do these without counters.

**K** $3 + 8 + 7$

**L** $2 + 5 + 5 + 8$

**M** $4 + 9 + 6$

**N** $2 + 5 + 1 + 4$

**O** $6 + 1 + 2 + 4$

**P** $4 + 8 + 1 + 2 + 6$

**Q** $5 + 2 + 5 + 2 + 5 + 5$

**R** $3 + 9 + 7 + 2 + 1$

**S** $6 + 5 + 1 + 9 + 4$

**T** $3 + 8 + 3 + 7 + 2$

**U** $4 + 8 + 5 + 1 + 6 + 7 + 2 + 5 + 3 + 5 + 9$

# Drawing shapes 2

► Copy each of these shapes carefully onto squared paper.
Write down the name of each shape.

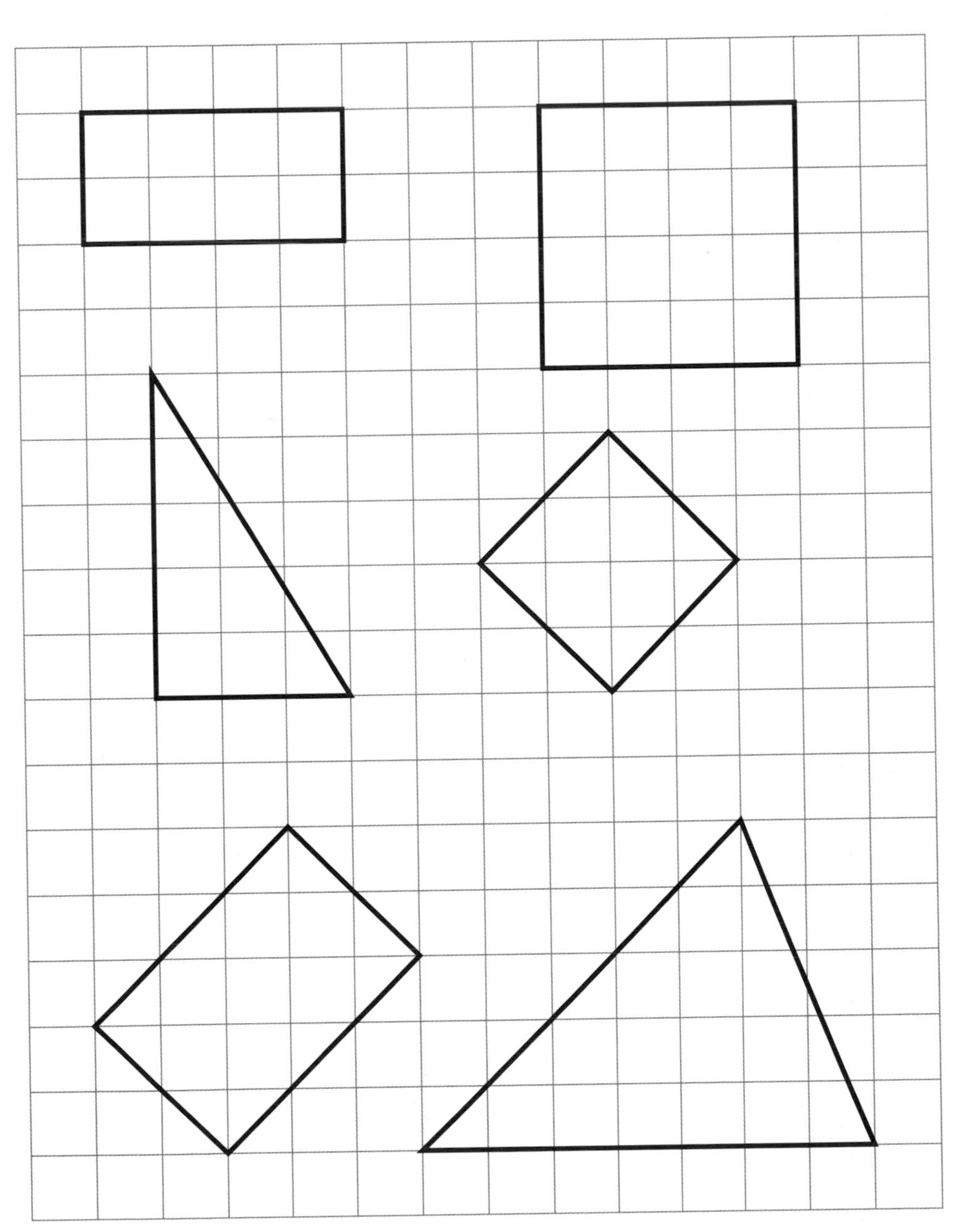

# Making tens

## Using cubes

These cubes are in **blocks of ten**.

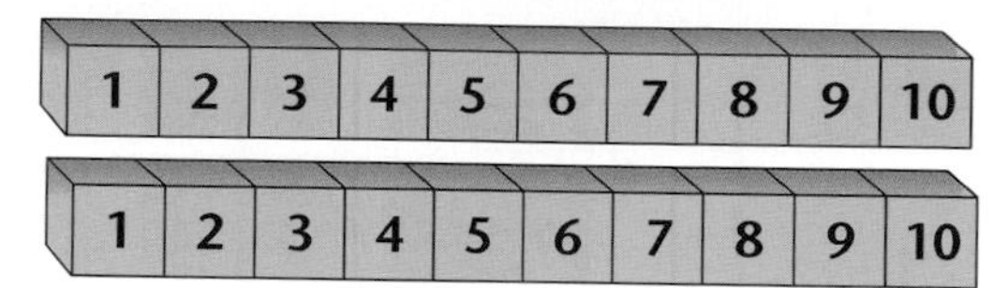


There are **20** cubes altogether.
10 + 10

These cubes are in **groups of ten**.

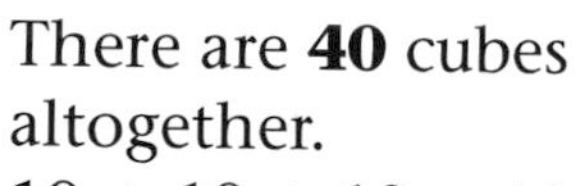

There are **40** cubes altogether.
10 + 10 + 10 + 10

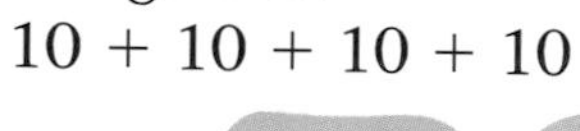

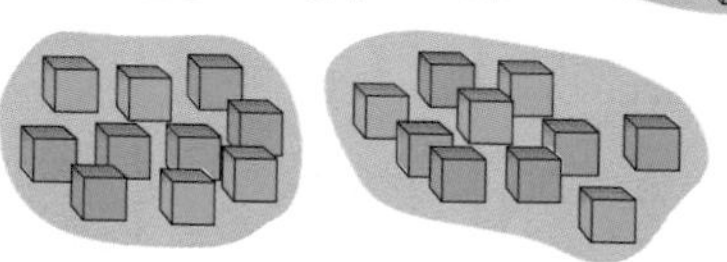

There are **35** cubes altogether in this picture.
10 + 10 + 10 + 5 more

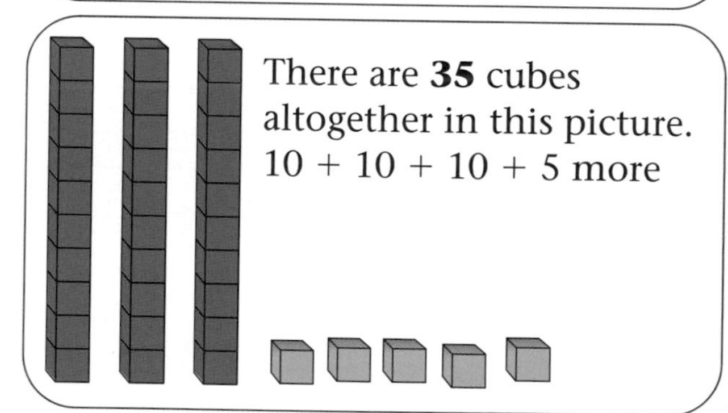

▶ You need the box of cubes.

Make these numbers using the cubes.
Show your blocks to your teacher.

**A** make 30

**B** make 60

**C** make 40

**D** make 32

**E** make 28

**F** make 15

10 ten

20 twenty

30 thirty

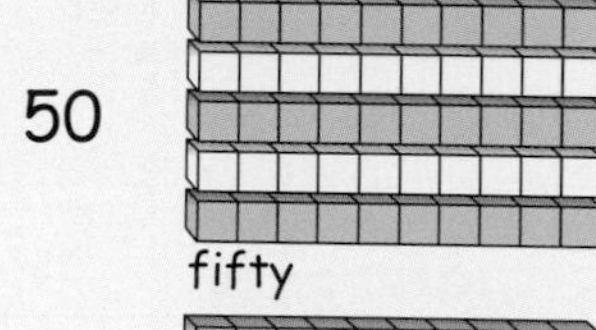

40 forty

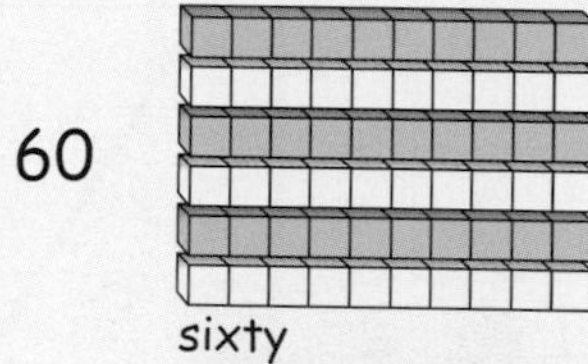

50 fifty

60 sixty

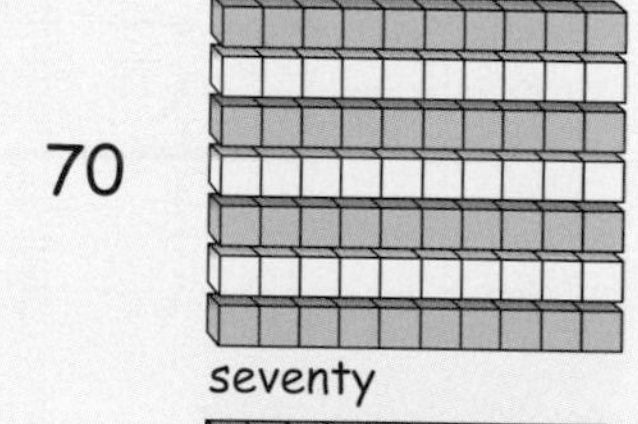

70 seventy

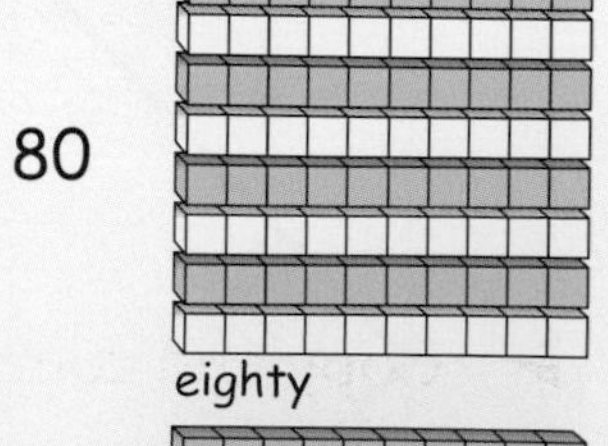

80 eighty

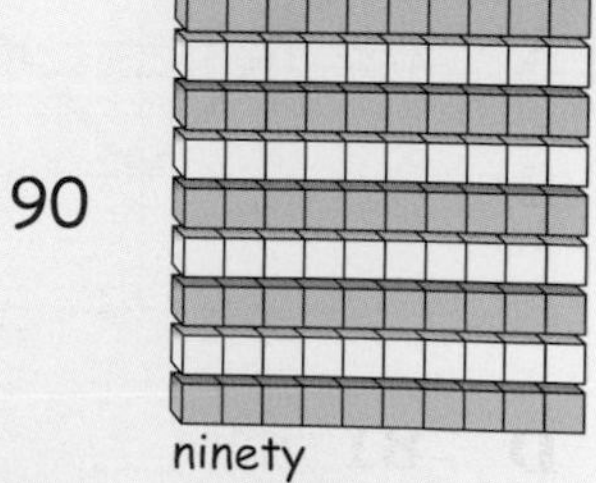

90 ninety

100 hundred

You need to practise counting in tens . . . or use worksheet 2/2.

## Place value

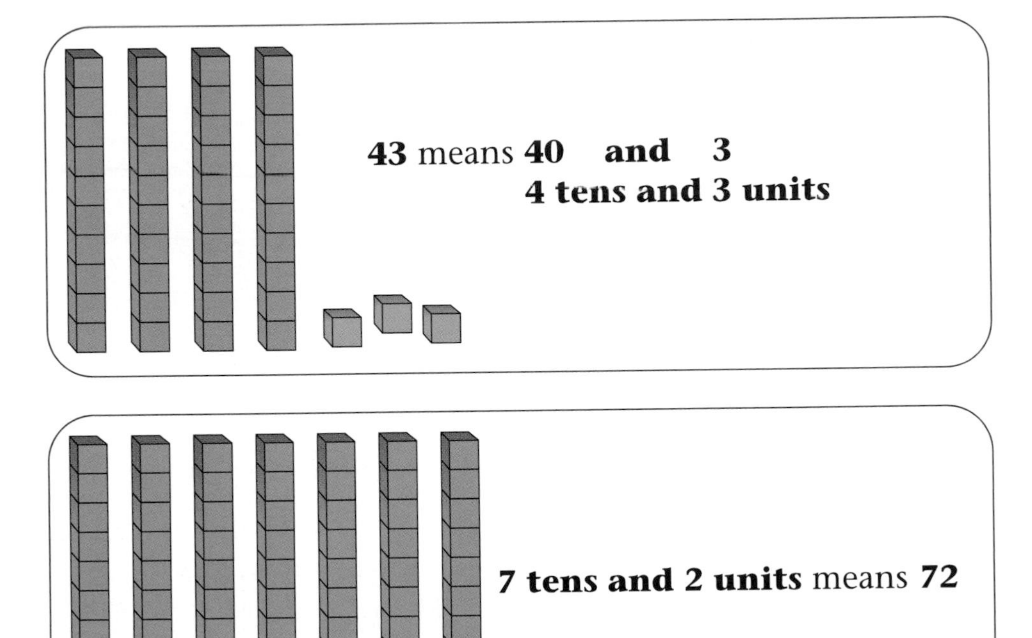


You need to practise using tens and units to write numbers . . . or use worksheet 2/3.

► Copy and complete these sentences.

**A** **62** has ____ tens and ____ units.

**B** **13** has ____ tens and ____ units.

**C** ____ has **4** tens and **9** units.

**D** **81** has ____ tens and ____ units.

**E** ____ has **2** tens and **7** units.

# Number patterns 1

4 2 7 4 2 7 4 2 7 4 . . .

This is a **repeating** number pattern. The next number is **2**.

You need to practise completing number patterns . . .
or use worksheet 2/4, section A.

3 4 5 6 7 8 9 10 11 12 . . .

This is **not** a repeating pattern.
This pattern goes **up** in **1s**. The next number is **13**.

2 5 8 11 14 . . .

This pattern goes
**up** in **3s**.
The next number is **17**.

70 60 50 40 . . .

This pattern goes
**down** in **10s**.
The next number is **30**.

You need to practise continuing number patterns . . .
or use worksheet 2/4, section B.

## Challenge

Can you make these number patterns?

**A** start at 4 . . . go up in 1s . . . finish at 10

**B** start at 20 . . . go up in 5s . . . stop after 6 numbers

**C** start at 50 . . . go down in 2s . . . stop after 10 numbers

**D** go up in 4s . . . stop after 5 numbers . . . finish at 3

# Unit 3

## Measuring units 1

| length | kilometre ↔ km | metre ↔ m |
|---|---|---|
| millimetre ↔ mm | centimetre ↔ cm | |

These are all units that are used for length. For shorthand, we use km, m, cm, mm.

▶ Copy these sentences using shorthand.
(. . . *or use worksheet 3/1*).

**A** Mount Everest is nearly **9 kilometres** high.

**B** A classroom door is about **2 metres** tall.

**C** A middle finger is about **1 centimetre** wide.

**D** The lead in a pencil is about **2 millimetres** across.

**E** A cricket pitch is **20 metres** long.

**F** A ping-pong ball is **38 millimetres** across.

**G** It takes about five minutes to walk **1 kilometre**.

**H** A ten-pin bowling lane is about **18 metres** long. Each pin is **38 centimetres** high.

You need to practise using shorthand for these units . . . or use worksheet 3/2.

### Challenge

What can you find in your house (or school) that is measured in millimetres, centimetres, metres, kilometres?

# Repeating patterns 2

This is a repeating pattern.

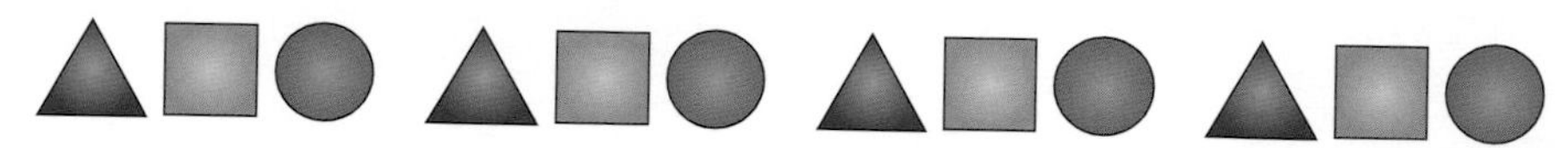

It has 4 repeats.

You need to practise completing shape patterns.
You need to practise counting the repeats in shape patterns
. . . or use worksheet 3/3.

Each part of the pattern has 3 objects.

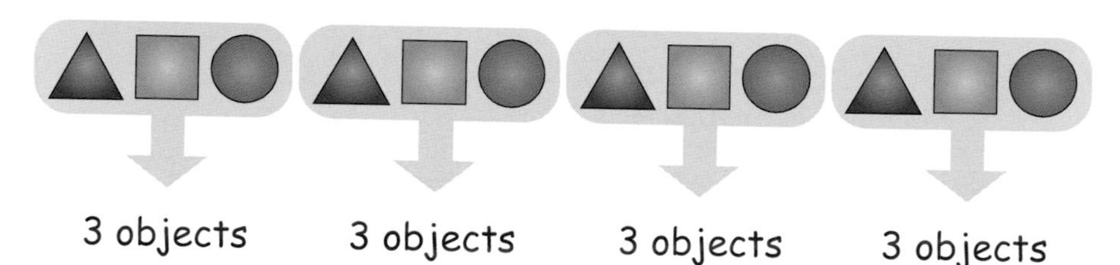


There are 4 objects in each part of this pattern.

▶ You need the cubes to answer these questions.

**1** **(a)** Make this pattern with your cubes.

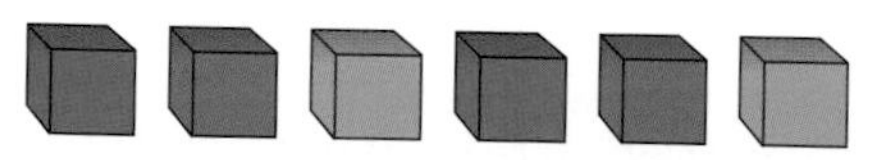

**(b)** Make another repeat of the pattern.

**(c)** Draw the pattern in your book like this.

**2** **(a)** Make a different repeating pattern with 3 cubes.

**(b)** Draw it in your book (or show it to your teacher).

**3** **(a)** Make this pattern with your cubes.

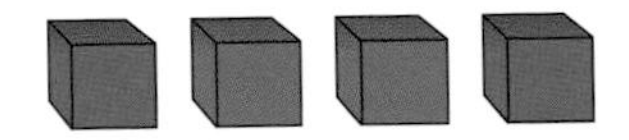

**(b)** Make **3** repeats of your pattern.

**(c)** Draw it in your book (or show it to your teacher).

**4** **(a)** Make a repeating pattern using **5** cubes.

**(b)** Draw it in your book (or show it to your teacher).

You need to practise counting the number of objects in each repeat . . . or use worksheet 3/4

## Digital clocks

This is a digital clock.

It shows the time **6:25**.

▶ Write down the time each clock shows.

**A**

**E**

**B**

**F**

**C**

**G**

**D**

**H**

# Add and subtract 3

## Adding

These all mean add

+ plus add sum total

**A** What is 4 add 5?

**B** Find the sum of 6 and 3.

**C** What is the total of 2 and 7?

**D** Find the sum of 7 and 9.

**E** What is the total of 8 and 3?

**F** What is 6 add 6?

**G** What is 4 plus 9?

**H** What is the total of 5 and 6?

## Subtracting

These all mean subtract

– minus subtract take away difference

**I** What is 4 take away 1?

**J** Find the difference between 6 and 3.

**K** What is 5 subtract 2?

**L** Find the difference between 3 and 9.

**M** What is 7 subtract 4?

**N** What is 8 take away 6?

**O** What is 9 minus 5?

**P** Find the difference between 2 and 7.

## Mixed problems

**A** Gary buys a pack of 8 invitations. He writes 6 of them. How many does he have left?

**B** A minibus has 6 passengers. 5 more get on. How many are there altogether?

**C** Jade warms up for 2 minutes then runs for 9 minutes. For how long has she exercised altogether?

**D** Ian drives 8 kilometres to work. He has driven 5 so far. How much further does he have to go?

**E** Anil buys a bag of 9 doughnuts. He eats 4 of them. How many does he have left?

**F** Natalie swims 5 lengths. She rests, then swims another 3 lengths.
How many lengths has she swum altogether?

### Dice

At the school fair you need to score a total of 8 with 2 dice to win the game.

**SCORE 8 TO WIN**

**1** Who has won a prize?

**2** Sue has scored 6 with her first dice. What does she need to score with the second dice to win a prize?

# Money 2

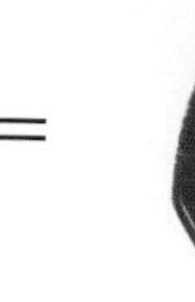

▶ Use real coins to help you with these questions.

**A** How many  =  ?

**B** How many  =  ?

**C** How many  =  ?

**D** How many  =  ?

**E** How many  =  ?

▶ Make these amounts of money from the coins above. Write down which coins you use each time.

| | | | | | | | |
|---|---|---|---|---|---|---|---|
| **F** | 70p | **I** | 40p | **L** | 80p | **O** | 90p |
| **G** | 45p | **J** | 35p | **M** | 57p | **P** | 26p |
| **H** | 63p | **K** | 34p | **N** | 28p | **Q** | 88p |

**R** Find the total amount of money in the picture.

## Challenge

**1** How many different ways can you find to make 50p?

What is the **smallest** number of coins you can use?
What is the **largest** number of coins you can use?

**2** How many different ways can you find to make £1?

What is the **smallest** number of coins you can use?

# Timetables

This timetable shows the times of a bus on the number 52 route.

| Town Centre | Riverside Park | Hilltop School | Church Square | Woodland Green | Bus Station |
|---|---|---|---|---|---|
| 10:00 | 10:11 | 10:24 | 10:38 | 10:42 | 10:50 |

The bus leaves the town centre at 10:00.
It arrives at Riverside Park at 10:11.

▶ Answer these questions.

**1** Look at the timetable above.

**(a)** What time does the bus get to Church Square?

**(b)** What time does the bus get to the bus station?

**(c)** Where is the bus at 10:24?

**2** This timetable shows the times of a train from Aden to Kimble.

| Aden | Borville | Dunton | Hayden | Kimble |
|---|---|---|---|---|
| 7:48 | 7:55 | 8:00 | 8:06 | 8:15 |

**(a)** What time does the train leave Dunton?

**(b)** Where is the train at 7:48?

**(c)** What time does the train get to Kimble?

**(d)** Where is the train at 8:06?

**3** This timetable shows the times of a coach from Birmingham to York.

| Birmingham | Nottingham | Leeds | York |
|---|---|---|---|
| 2:15 | 3:45 | 5:50 | 7:05 |

**(a)** What time does the coach leave Birmingham?

**(b)** When does the coach get to Leeds?

**(c)** What time does the coach arrive at York?

**(d)** Where is the coach at 3:45?

**4** This is the timetable for Easton School.

**(a)** When does tutor time start?

**(b)** What lesson starts at 1:00?

**(c)** When does lunch start?

**(d)** What happens at 3:30?

| | |
|---|---|
| Tutor time | 9:00 |
| Lesson 1 | 9:15 |
| Break 1 | 10:25 |
| Lesson 2 | 10:45 |
| Lunch | 11:55 |
| Lesson 3 | 1:00 |
| Break 2 | 2:10 |
| Lesson 4 | 2:20 |
| Home time | 3:30 |

## Challenge

Make a timetable for your school day, or a bus route near your home.

# Solid shapes 1

*you may need the solids kit*

| cube | pyramid | cone | sphere |
| --- | --- | --- | --- |
| | | | |

▶ Write down the name of each solid.

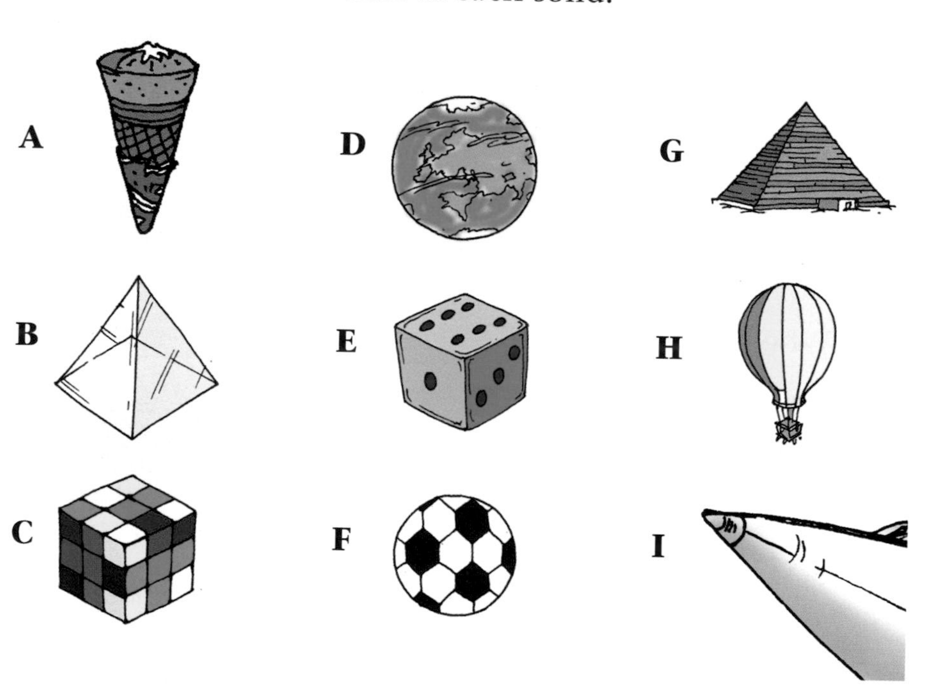


## Challenge

Find objects at home or at school that are solid shapes.
Try to find a cube . . . a pyramid . . . a cone . . . a sphere.

# Unit 4

## Sorting 2

**1** These items have been sorted into 2 groups.

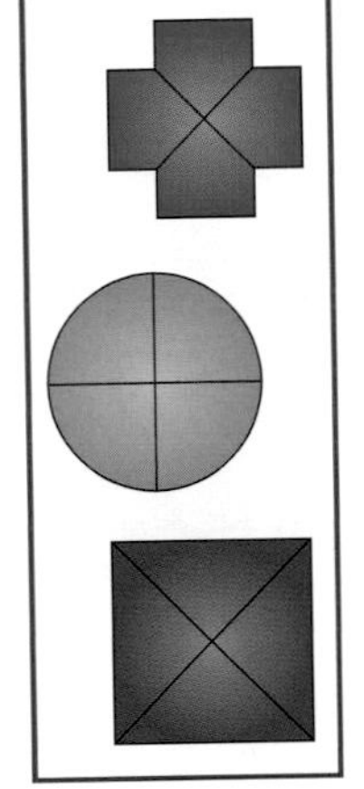

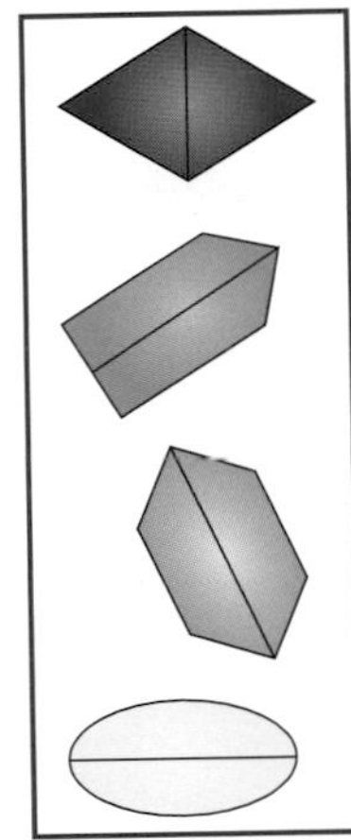

**(a)** Which group will this item go into? Why?

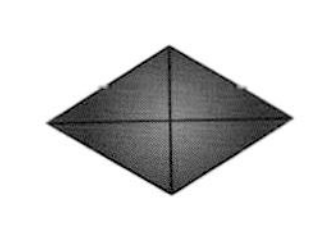

**(b)** Which group will this item go into? Why?

**2** These items have been sorted into 3 groups.

**(a)** Which group will this item go into?

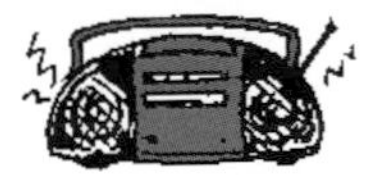

**(b)** Give a reason.

**3** These groups need names.

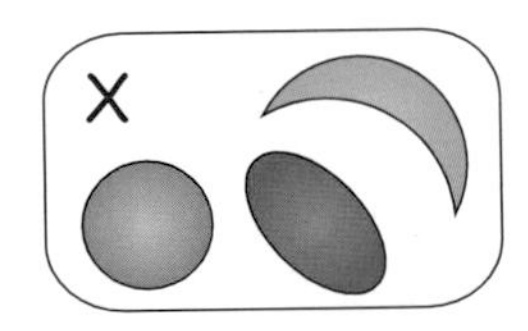


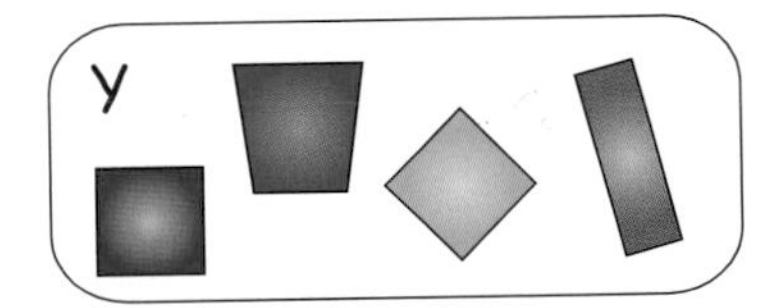

**(a)** Write down a name (or sentence) for each group.

**(b)** Which group will this item go into?

**(c)** Which group will this item go into?

You need to practise sorting objects into groups . . . or use worksheet 4/1.

# Reading scales 2

## Review

*Look back at page 18 if you cannot answer this question easily.*

▶ Write down the number each arrow points to.

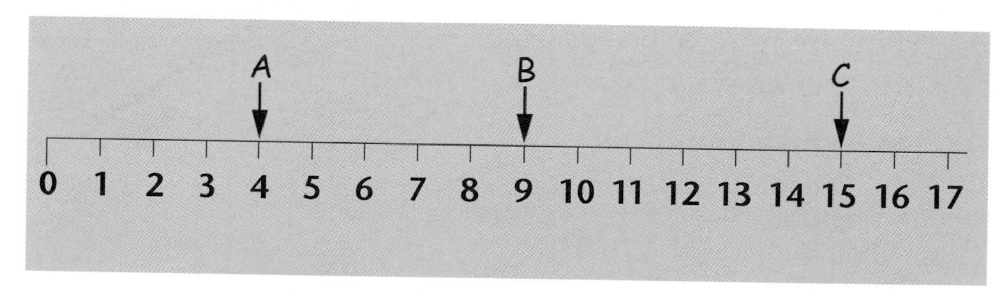


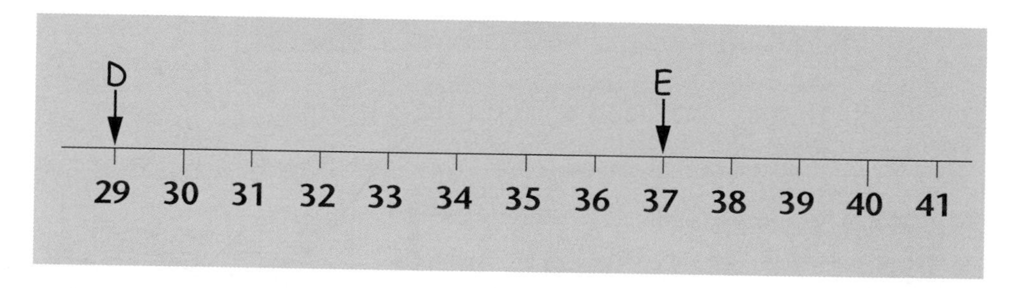


You need to practise using number lines . . . or use worksheet 4/2, section A.

## Missing numbers

On these scales, only the **tens** numbers are labelled.

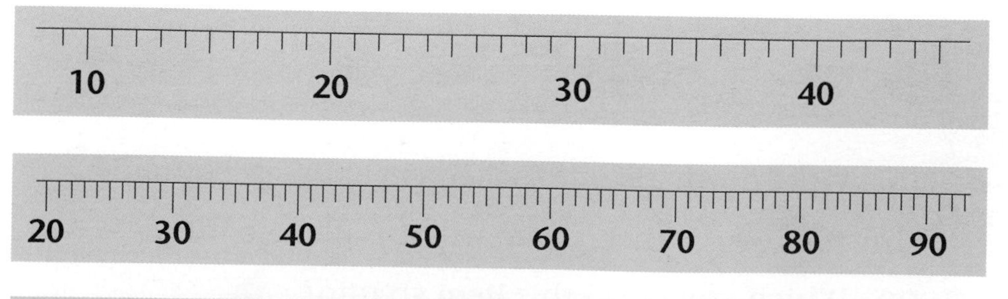


You need to practise marking numbers on scales . . . or use worksheet 4/2, section B.

## Reading scales

This mark is one past the ten so it is worth 10 + 1 = **11**

This mark is two past the twenty so it is worth 20 + 2 = **22**

This mark is five past the thirty so it is worth 30 + 5 = **35**

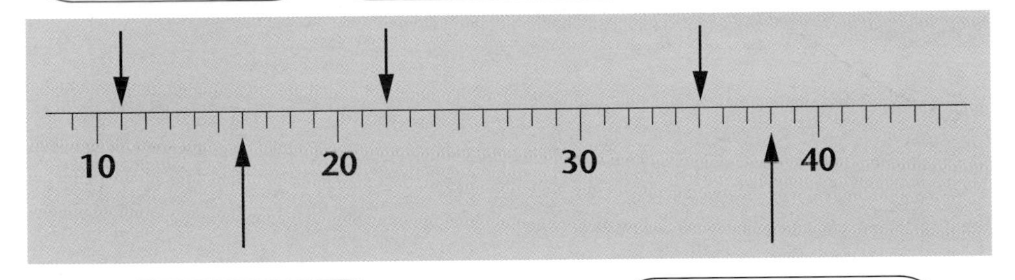


This mark is six past the ten so it worth 10 + 6 = **16**

This mark is eight past the thirty so it is worth 30 + 8 = **38**

► Write down the number each arrow points to.

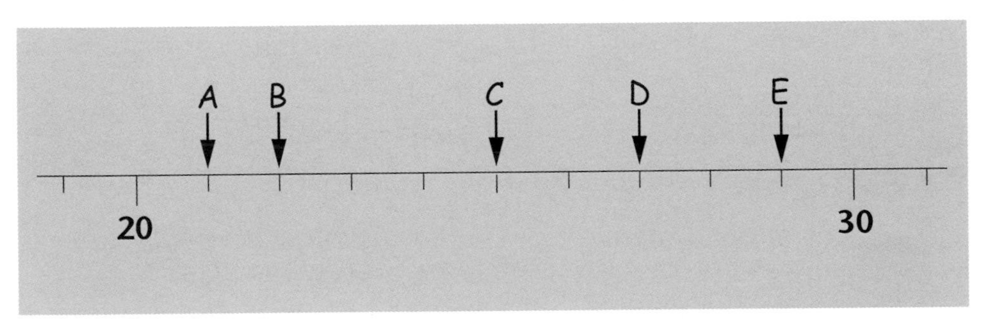


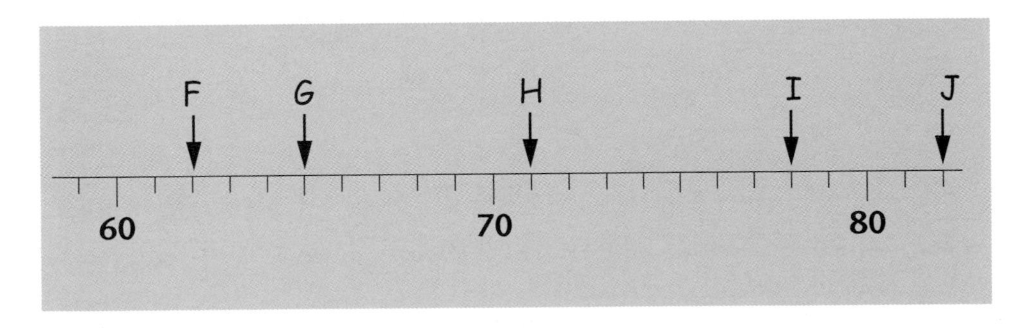

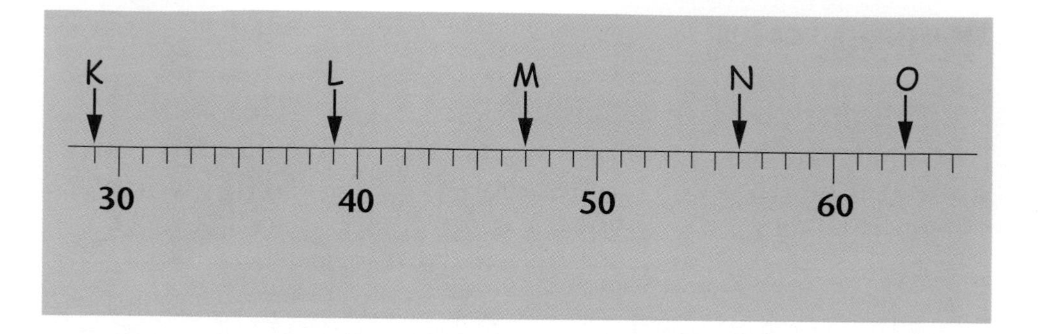


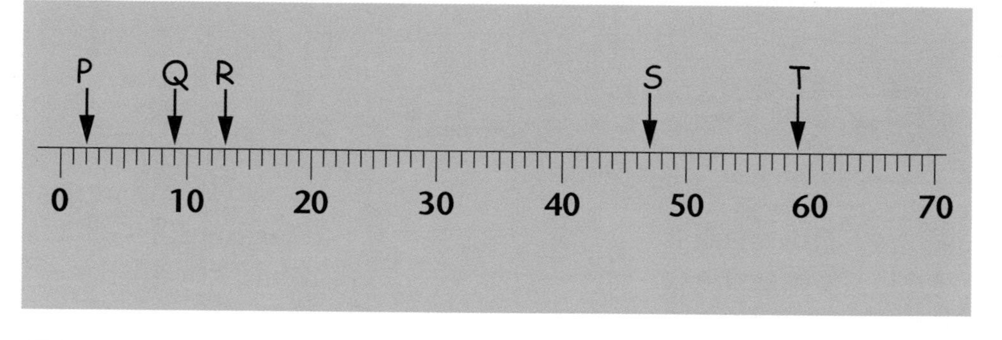


You need to practise matching numbers with scales . . . or use worksheet 4/2, section C.

## Challenge

This is a family number line.

The ages of each member of the family have been marked on the line with arrows.

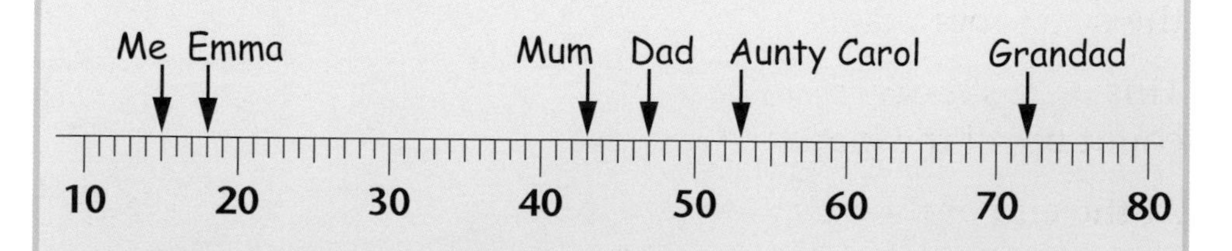


► Make a number line marked in tens. (You could use a piece of wallpaper.)

Put arrows on it to show the ages of members of *your* family.

# Tallies 1

Susie works in a clothes shop.

She must find out how many people bring a dress into the changing room today but do not buy it.

She is keeping a **tally**.

The first time it happens she puts a mark.

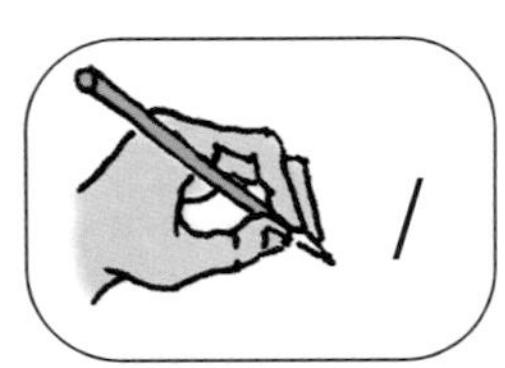


and the next . . . and the next . . . and the next . . .

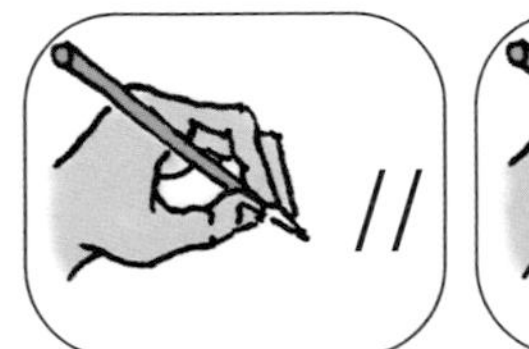


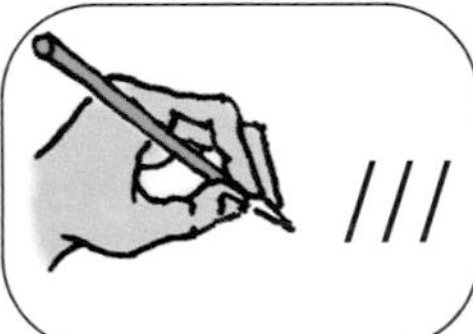


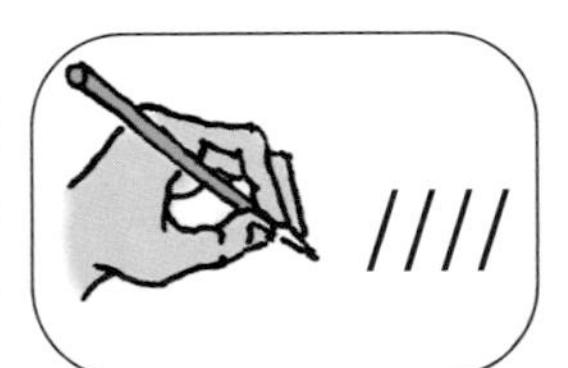


When she gets to **five** she puts a line across the other four marks.

This makes it easier to count up all the marks.

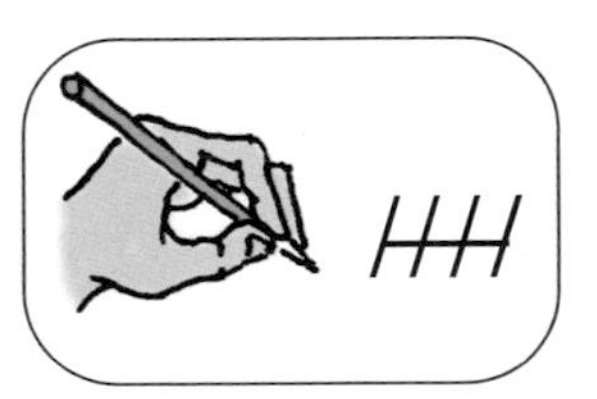


At the end of the day her paper looks like this.

There are **12** marks altogether.

| | |
|---|---|
| 1 | / |
| 2 | // |
| 3 | /// |
| 4 | //// |
| 5 | ~~////~~ |
| 6 | ~~////~~ / |
| 7 | ~~////~~ // |
| 8 | ~~////~~ /// |
| 9 | ~~////~~ //// |
| 10 | ~~////~~ ~~////~~ |
| 11 | ~~////~~ ~~////~~ / |
| 12 | ~~////~~ ~~////~~ // |
| 13 | ~~////~~ ~~////~~ /// |
| 14 | ~~////~~ ~~////~~ //// |
| 15 | ~~////~~ ~~////~~ ~~////~~ |
| 16 | ~~////~~ ~~////~~ ~~////~~ / |
| 17 | ~~////~~ ~~////~~ ~~////~~ // |

You need to practise tallying objects . . . or use worksheet 4/3.

# Comparing

## Comparing with words

Words can be used to compare numbers.

| how many | more | less | the same as |
|---|---|---|---|
| greater than | fewer | more than | less than |

Bill has 6 sisters and 2 brothers.

He has **more** sisters than brothers.

He has **fewer** brothers than sisters.

He has **4 more** sisters than brothers.

He has **4 less** brothers than sisters.

28 is **bigger** than 25 . . . . . . . . 25 is **smaller** than 28

28 is **greater** than 25 . . . . . . . 25 is **less** than 28

28 is **three more** than 25 . . . . 25 is **three less** than 28

▶ Copy and complete these sentences.

Use the words at the top of the page.

**A** 3 is ____________ than 7.

**B** 85 is ____________ than 14.

**C** 9 is ____________ nine.

**D** 26 is ____________ than 62.

**E** 98 is ____________ than 89.

**F** 21 is ____________ than 20.

**G** 0 is ____________ than 3.

## Comparing with numbers

▶ Find the missing number in each sentence.

**A** 10 is ____ more than 6.

**B** 20 is ____ less than 25.

**C** 18 is ____ greater than 15.

**D** 58 is ____ fewer than 68.

**E** 14 is ____ more than 1.

**F** 32 is ____ greater than 12.

**G** 9 is ____ less than 13.

## Comparing problems

**1** Rob has 8 new computer games and 15 borrowed games.
Which does he have more of?

**2** Sue has 47 CDs and 39 singles.
How many more CDs does she have than singles?

**3** Mia posts 13 letters and 6 parcels.
Which is greater, the number of letters posted or the number of parcels?

**4** Gregor weighed 85 kilograms before his diet.
Now he weighs 78 kilograms.
How many kilograms less does he weigh?

**5** Pat buys 5 chocolate cakes, 2 cans of lemonade, 5 cream cakes and 7 cans of cola.
Complete the sentences.

The number of cans of lemonade is _____ the number of cans of cola.

The number of chocolate cakes is _____ the number of cream cakes.

The total number of cakes is _____ the total number of cans of drink.

# Change 1

## Change from 50p

How much change do you get from 50p if you spend 30p?

spend　　　　　　　　　　　　change

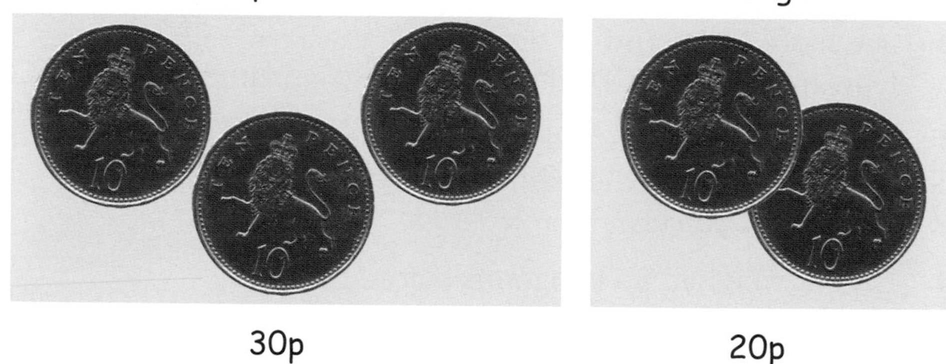

30p　　　　　　　　　　　　20p

You get 20p change.

▶ Work out how much change you get from 50p if you spend these amounts.
Use coins to help you.

**A** 10p　　　　　　**C** 40p

**B** 20p　　　　　　**D** 50p

## Change from £1

How much change do you get from £1 if you spend 80p?

spend | change

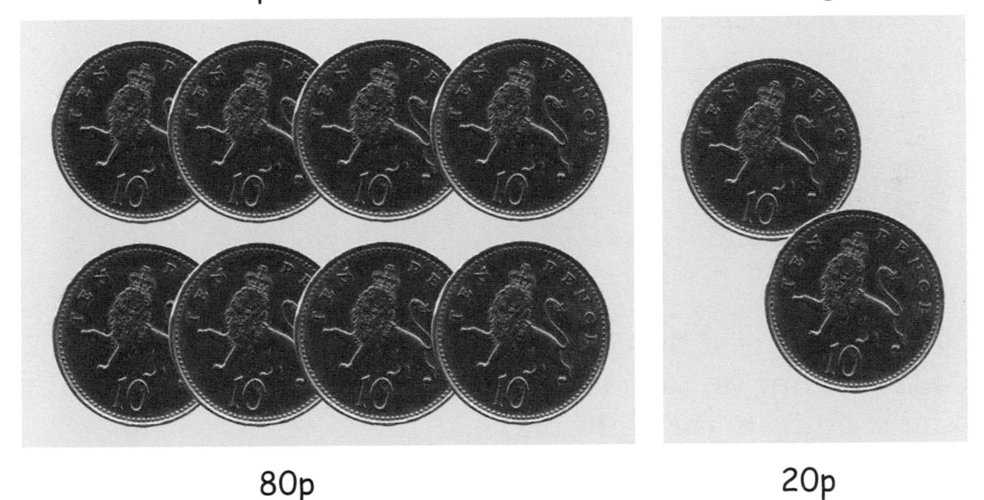

80p | 20p

You get 20p change.

▶ Work out how much change you get from £1 if you spend these amounts.
Use coins to help you.

**E** 90p
**F** 60p
**G** 40p
**H** 20p
**I** 30p
**J** 10p
**K** 50p
**L** 70p

# Left and right

## Hands

This is your **left** hand. 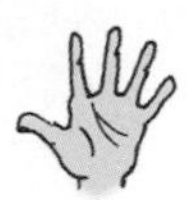 This is your **right** hand. 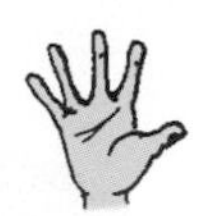

▶ Which hand is in each picture?

A 

B 

C 

D 

E 

## The cinema

Five friends are at the cinema.
Mike is sitting on Sue's **left**.
Kim is sitting to the **right** of Dan.

▶ Copy and complete these sentences.

**F** Anika is sitting on Dan's ______ .

**G** Sue is sitting on Mike's ______ .

**H** Sue is sitting to the ______ of Anika.

**I** Anika is sitting to the left of ______ and to the right of ______ .

**J** Dan is sitting to the right of ______ and to the left of ______ .

## Dog rescue

These dogs are waiting to be fed.

Toby is at the left hand side of the picture, but be careful!

Toby is on **Growler's right**.

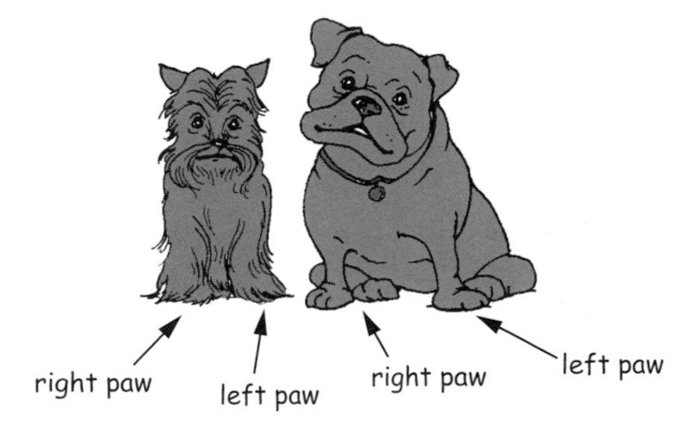


▶ Copy and complete these sentences.

**K** Eddie is on Growler's ______ .

**L** Bouncer is on Chaser's ______ .

**M** Sam is to the ______ of Chaser.

**N** Eddie is to the ______ of Growler and to the ______ of Sam.

**O** Sam is to the left of ______ and to the right of ______ .

### Challenge

Make up some sentences of your own like these.

They could be about people in your class, or your family, or a picture of your favourite group . . .

# Unit 5

## Comparing weights

heavier    lighter    heaviest    lightest

This bag of sugar weighs 1 kg.

Remember: 1 kg means 1 kilogram.

SUGAR

This banana weighs **less than** a bag of sugar.

This banana is **lighter than** 1 kg.

COLA

This bottle of cola weighs **more than** a bag of sugar.

This bottle of cola is **heavier than** 1 kg.

► Are these heavier or lighter than 1 kg?
For each one write down heavier than 1 kg or lighter than 1 kg.

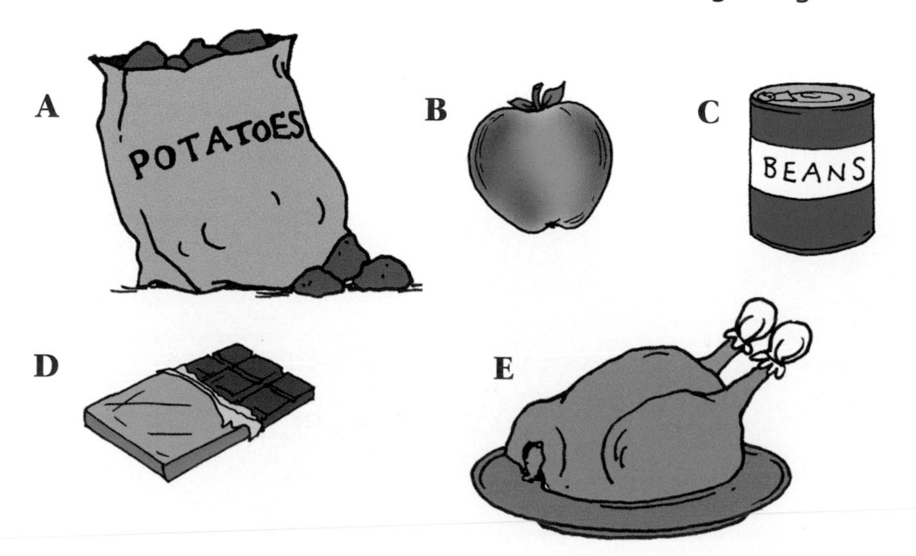

▶ Answer these questions.

1 **(a)** Which is heavier than the man, the elephant or the dog?

**(b)** Which is lighter than the man?

2 **(a)** Which is heavier than the car, the bicycle or the bus?

**(b)** Which is lighter than the car?

3 **(a)** Which is the heaviest?

**(b)** Which is the lightest?

**(c)** Write the creatures in order of weight.

Start with the lightest.

squirrel bird mouse bee cat

# Adding 3

*you may use the 100 line*

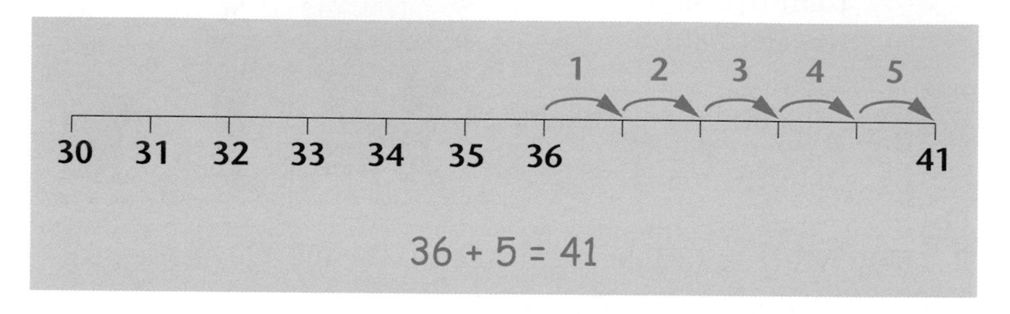


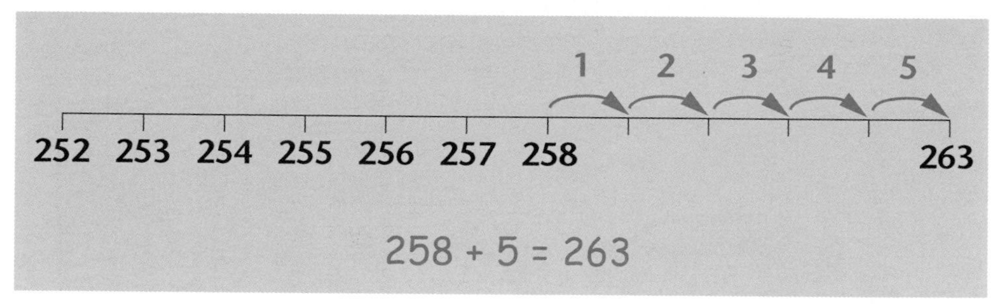


► Work these out without a calculator.

**1** 15 + 3

**2** 24 + 5

**3** 36 + 3

**4** 7 + 51

**5** 62 + 6

**6** 47 + 2

**7** 8 + 60

**8** 34 + 7

**9** 56 + 7

**10** 8 + 25

**11** 49 + 5

**12** 87 + 5

**13** 13 + 8

**14** 9 + 94

**15** 124 + 3

**16** 256 + 2

**17** 482 + 7

**18** 855 + 1

**19** 303 + 5

**20** 521 + 8

**21** 7 + 233

**22** 6 + 604

**23** 378 + 4

**24** 222 + 9

**25** 766 + 6

**26** 983 + 8

**27** 7 + 439

**28** 806 + 5

**29** 197 + 4

**30** 8 + 685

# Measuring units 2

kilogram ↔ kg  gram ↔ g  litre ↔ l  millilitre ↔ ml

Grams and kilograms are units that are used for **weight** (how heavy things are).

Litres and millilitres are units that are used to measure amounts like water in a jug.

For shorthand, we use kg, g, l, ml.

▶ Copy these sentences using shorthand.
(. . . *or use worksheet 5/1*).

**A** A bag of sugar weighs **1 kilogram**.

**B** A football weighs about **400 grams**.

**C** A kettle holds about **2 litres**.

**D** A teaspoon holds **5 millilitres**.

**E** An adult should eat **18 grams** of fibre a day.

**F** An adult should drink at least **2 litres** of water a day.

**G** There are 100 calories in a cup of **250 millilitres** of skimmed milk.

**H** An adult female weighing **55 kilograms** should eat **29 grams** of protein a day.

 You need to practise using shorthand for these units . . . or use worksheet 5/2.

## Challenge

What can you find in your house (or school) that is measured in kilograms, grams, litres, millilitres?

# Repeating patterns 3

This is a **repeating** pattern.
It is made using two shapes.

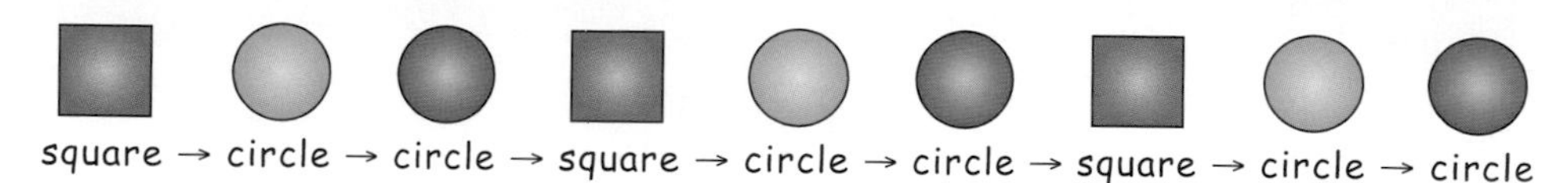


Repeating patterns are used in wallpaper and tiling patterns.

► Which of these patterns are **repeating** patterns?
Write *yes* or *no*.

A

B

C 

D **MHIMHIMHIWHIMHIMHI**

E 

You need to practise continuing shape patterns . . . or use worksheet 5/3.

This is a repeating pattern.

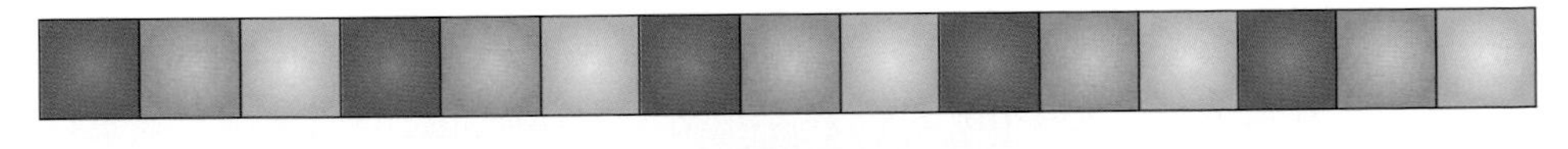

This is part of the pattern. 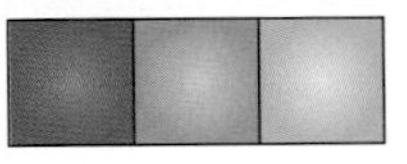

This is a different part of the pattern. 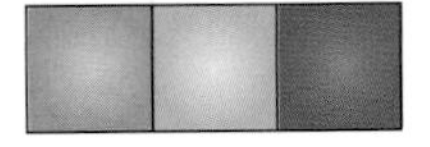

One of these is another part of the pattern.
Check that you know which one is correct.

**A** 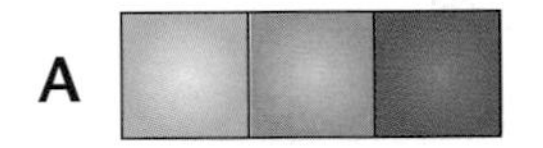 **B** 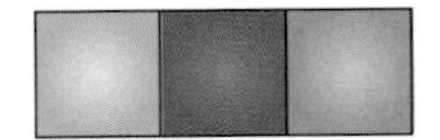 **C** 

► Each block belongs to one of the repeating patterns.
Match them up.

pattern A

pattern B

pattern C

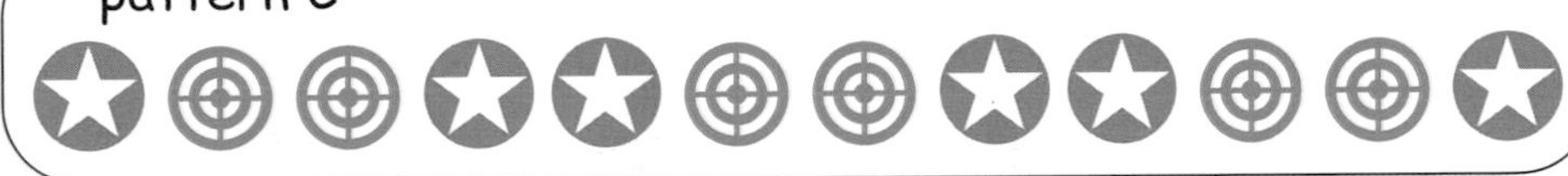

pattern D

**1**  **2**  **3** 

**4**  **5**  **6** 

# Two times table

lots of   multiply   ×   times   double   twice

Ben makes earrings.

He puts two earrings in each pack.

► Copy and complete the pattern.

1 lot of 2 = 2 

2 lots of 2 = 4 

3 lots of 2 = ☐ 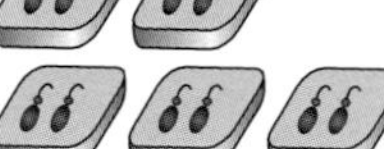

4 lots of 2 = ☐

5 lots of ☐ = 10

6 ☐ 2 = 12

7 lots of ☐ = ☐

☐ lots of ☐ = ☐

☐ lots of ☐ = ☐

10 lots of ☐ = ☐

$1 \times 2 = 2$

$2 \times 2 = 4$

$3 \times 2 = 6$

$4 \times 2 = 8$

$5 \times 2 = 10$

$6 \times 2 = 12$

$7 \times 2 = 14$

$8 \times 2 = 16$

$9 \times 2 = 18$

$10 \times 2 = 20$

This is another way to show the two times table.

| 1 |
|---|
| **2** |

| 1 | 3 |
|---|---|
| 2 | **4** |

| 1 | 3 | 5 |
|---|---|---|
| 2 | 4 | **6** |

| 1 | 3 | 5 | 7 |
|---|---|---|---|
| 2 | 4 | 6 | **8** |

| 1 | 3 | 5 | 7 | 9 |
|---|---|---|---|---|
| 2 | 4 | 6 | 8 | **10** |

► Can you draw the shape for $7 \times 2$?

▶ Find the answers.

**A** $4 \times 2 =$

**B** $9 \times 2 =$

**C** $3 \times 2 =$

**D** $7 \times 2 =$

**E** ✣ $\times 2 = 10$

**F** ❁ $\times 2 = 16$

**G** ✳ $\times 2 = 12$

**H** ★ $\times 2 = 2$

**I** double five

**J** two lots of two

**K** twice eight

**L** three multiplied by two

**M** double nine

**N** ___ multiplied by two = twelve

**O** one times two

**P** twice ten

# Writing numbers 1

▶ Copy and complete these sentences.

**A** **52** is **50** + **2** = **fifty-two**

**B** **38** is **30** + **8** = ________

**C** **93** is __ + __ = ________

**D** **27** is __ + __ = ________

**E** **45** is __ + __ = ________

You may need the Question and Answer cards.

# Right angles 1

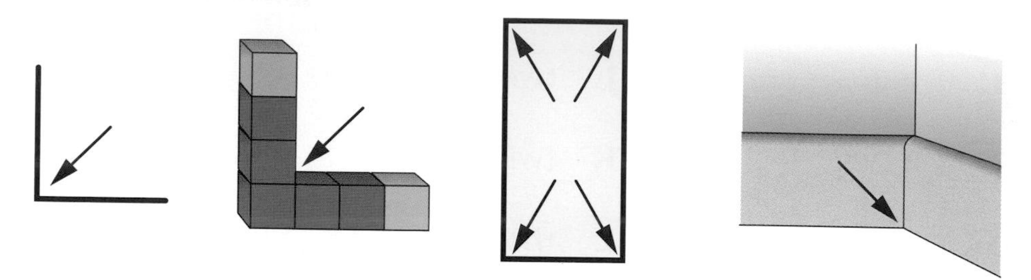

All of the pictures above show right angles.
A right angle is a corner which measures 90 degrees.
A right angle is a quarter of a turn.

▶ Which of these angles are right angles?
Check with the corner of a piece of paper or a right angle made from cubes.

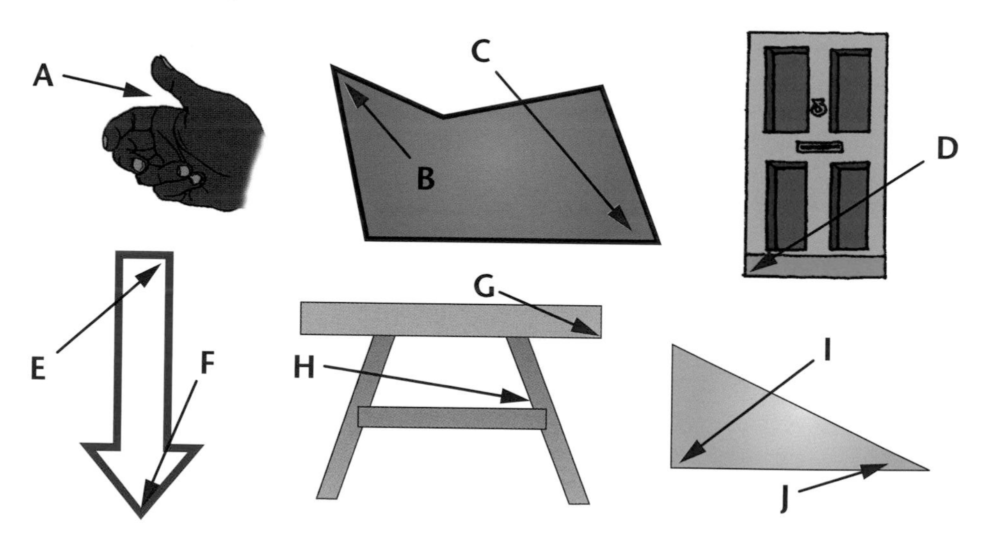


## Challenge

Find as many right angles as you can . . . in the classroom
. . . at home.

Can you describe where they are?

# Unit 6

## Shapes 3

| triangle | square | pentagon | hexagon | octagon |
|---|---|---|---|---|
| | | | | |

▶ Match up the ends of the sentences and copy them out.

| | |
|---|---|
| A pentagon has | 8 sides, 8 corners |
| A hexagon has | 5 sides, 5 corners |
| An octagon has | 6 sides, 6 corners |

▶ Write down the name of each shape.

**A** 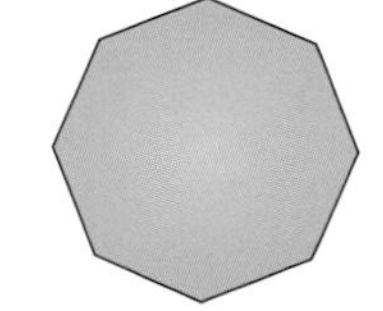

**B** 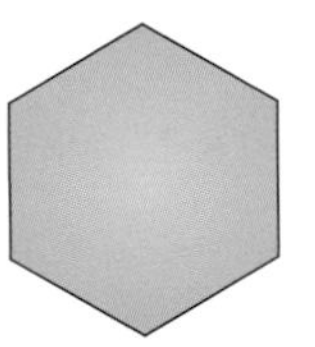

**C** 

**D** 

**E** 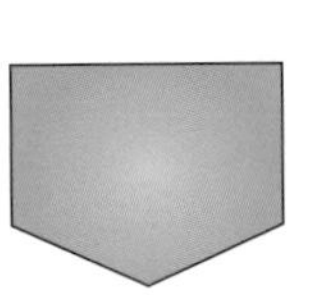

**F** 

**G** 

**H** 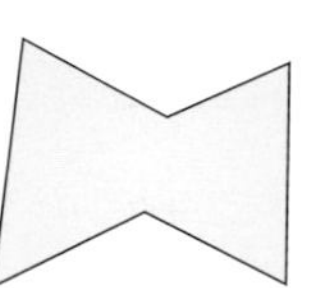

**I** 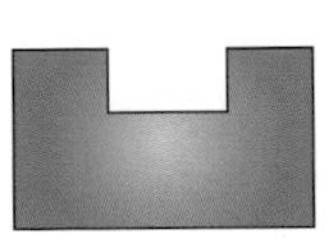

**J** 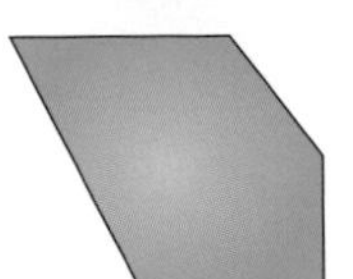

**K** 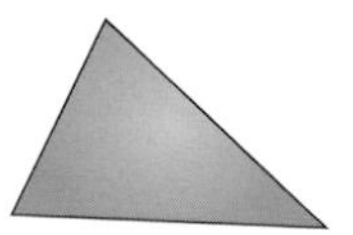

**L** 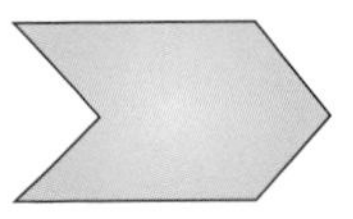

# Using a calculator 1

## Numbers on a calculator

To enter 47 into a calculator, first press the 4 button.

Now press the 7 button.

| Number | Calculator |
|---|---|
| 0 | 0 |
| 1 | 1 |
| 2 | 2 |
| 3 | 3 |
| 4 | 4 |
| 5 | 5 |
| 6 | 6 |
| 7 | 7 |
| 8 | 8 |
| 9 | 9 |

► Use your calculator to work these out.

**A** 47 + 21
Press:

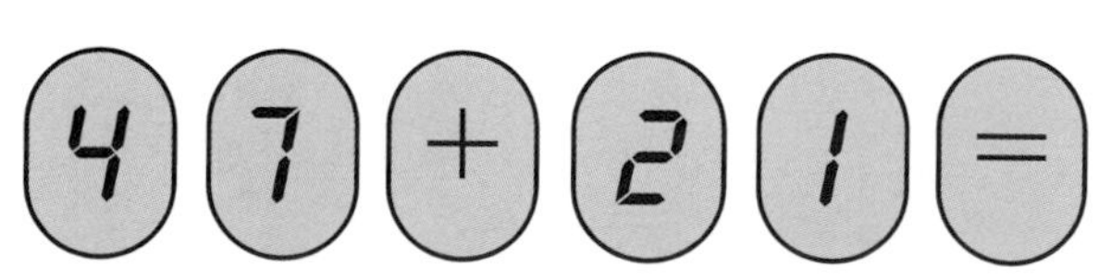


**B** 36 + 19

**C** 55 + 43

**D** 27 + 91

**E** 78 + 34

**F** 85 − 23

**G** 92 − 19

**H** 57 − 38

**I** 64 − 48

**J** 75 − 26

**K** 46 + 34

**L** 37 + 98

**M** 65 − 21

**N** 72 – 55

**O** 59 + 49

▶ Use your calculator to work these out.
Write down the sum you enter into the calculator.

**P** Thirty-seven add twenty-five.

**Q** The sum of forty-two and fifty-one.

**R** Seventy-four subtract fifteen.

**S** Sixty-nine take away eighty.

**T** The difference between forty-five and twenty-seven.

**U** The total of ninety-one and nineteen.

**V** Twenty-eight add sixty.

**W** The difference between eighty-one and sixteen.

**X** The sum of forty-seven and eight.

**Y** Seventy-one subtract thirty-eight.

## Calculator problems

▶ Work out the answers to these problems. Write down the sum that you do.

**A** Jack buys a T-shirt for £15 and a pair of jeans for £34.
How much does he spend altogether?

**B** There are 42 people on a bus.
At the next stop another 11 people get on.
How many people are on the bus now?

**C** Fran has saved up £57.
On her birthday she gets another £25.
How much money does she have now?

**D** 

Darren cycles 21 kilometres.
He rests, then cycles 34 kilometres more.
How far has he cycled altogether?

**E** Shannon has 81 CDs.
She gets 12 more.
How many CDs does she have now?

# Times of the day

morning afternoon evening
midday midnight

10 **am** is **morning**
3 **pm** is **afternoon**
8 **pm** is **evening**

► Choose the best word for each time.
morning, afternoon or evening

**A**
doctors
9:30 am

**B**
lunch
12:45 pm

**C**
cinema
7:15 pm

**D**
shopping
2:00 pm

**E**
train home
8:40 pm

**F**
go to work
8:30 am

**G**
tea break
3:15 pm

**H**
ice skating
6:00 pm

| | |
|---|---|
| | 12:00 am |
| | 1:00 am |
| | 2:00 am |
| | 3:00 am |
| | 4:00 am |
| | 5:00 am |
| morning | 6:00 am |
| | 7:00 am |
| | 8:00 am |
| | 9:00 am |
| | 10:00 am |
| | 11:00 am |
| | 12:00 pm |
| | 1:00 pm |
| | 2:00 pm |
| afternoon | 3:00 pm |
| | 4:00 pm |
| | 5:00 pm |
| | 6:00 pm |
| | 7:00 pm |
| evening | 8:00 pm |
| | 9:00 pm |
| | 10:00 pm |
| | 11:00 pm |
| | 12:00 am |

▶ Answer these questions.

**1** Copy and complete this list. The first one has been done for you.

half past seven in the morning is 7:30 am

four o'clock in the afternoon is ______

ten past nine in the evening is ______

five to three in the morning is ______

twenty to eleven in the evening is ______

**2** Alan starts work at 8 pm.
Does he work during the **day** or the **night**?

**3** Shirley has an early morning swim at 6:30.
Is this 6:30 **am** or 6:30 **pm**?

**4** Midday is in the middle of the day. It is 12:00 pm.
What time is midnight?

**5** Marcus watches the breakfast news at six o'clock.
Write this time using am or pm.

**6** Stefan finishes school at 3:20.
Is this 3:20 am or 3:20 pm?

**7** 

Tina watches the sunset at quarter to eight.
Write this time using am or pm.

**8** Gary must get to the post office before midday.
Is this morning, afternoon or evening?

# Halves 1

## One half

Each of these shapes is split into **two equal parts**.
Each part is called **one half**.

Two equal parts are called **halves**.

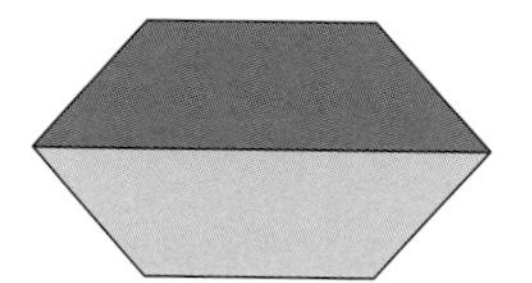
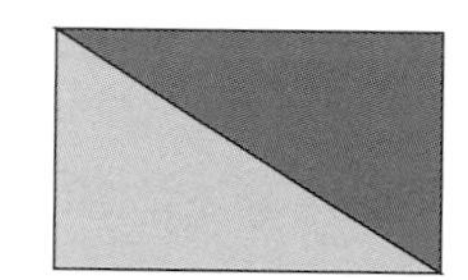

These shapes are **not** split in half. Can you see why?

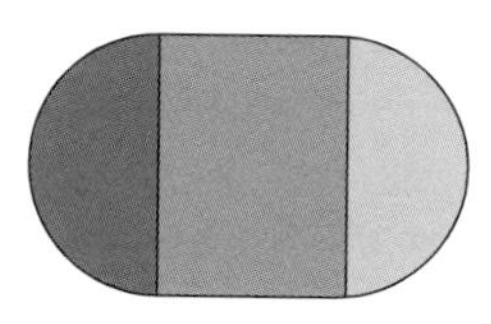
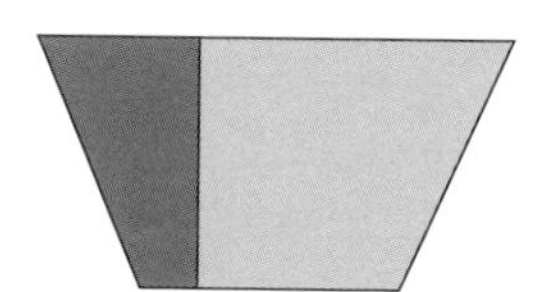

▶ This is Tim's homework.
Each shape should be split in half, but he has made some mistakes.
For each shape, say right or wrong.

**A**

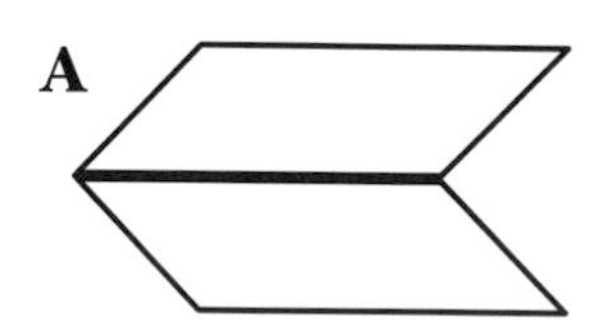

**B**

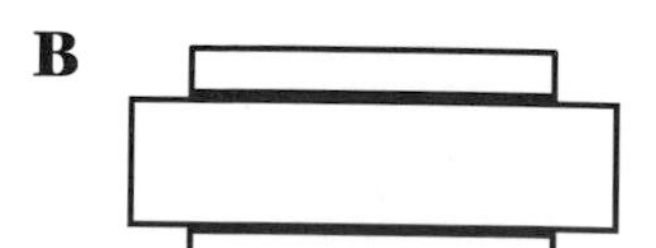

**C**

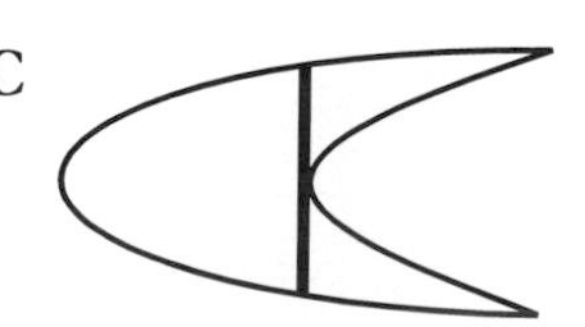

**D**

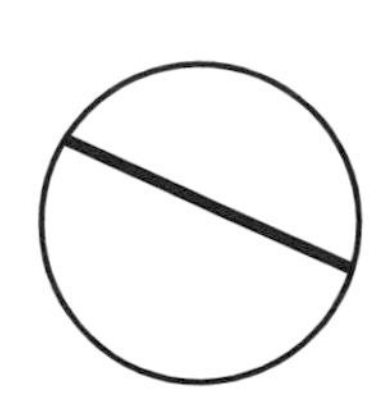

**E**

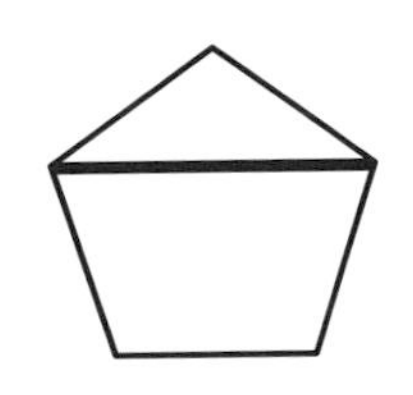

**F**

**G**

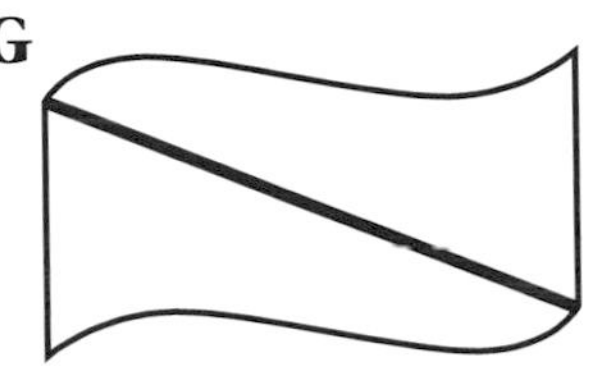

**H**

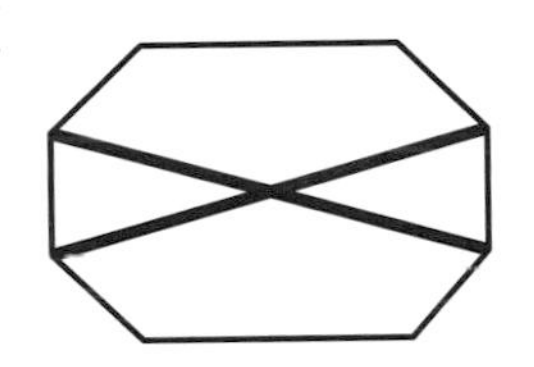

**I**

## Equal parts

▶ Is each shape split into equal parts?
Write yes or no.

**A**

**B**

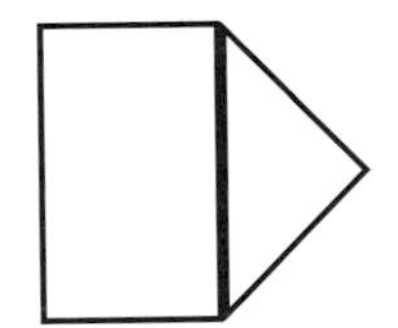

**C**

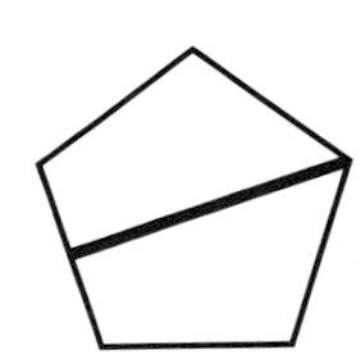

**D**

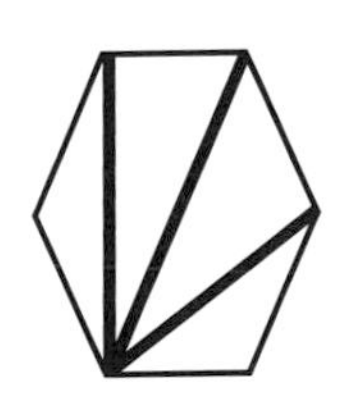

**E**

**F**

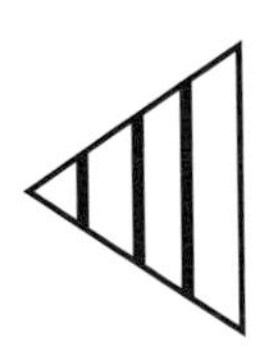

**G**

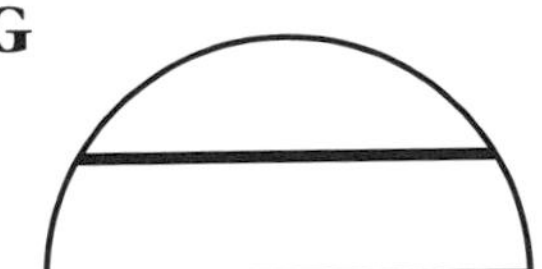

**H**

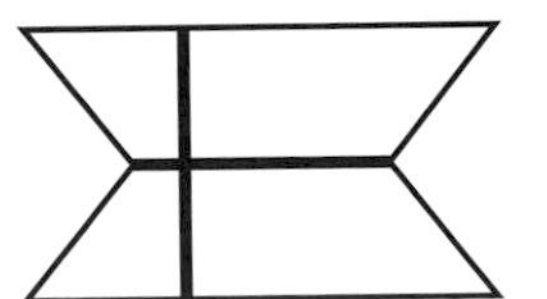

**I**

# Position 2

These people are in a queue.

Ann is **at the front** of the queue. She is **in front** of Bill.
Bill is **behind** Ann and **in front** of Chris.

▶ Copy and complete these sentences.

**A** Emma is in front of _______ .

**B** Dave is behind _______ .

**C** Gail is in front of _______ and behind _______ .

**D** Chris is _______ Dave and _______ Bill.

**E** _______ is at the back of the queue.

These are two shelves in a shop.

The salt is **above** the fish. The cat food is **below** the beans.
The tea bags are **next to** the salt and **above** the soap.

**F** The beans are above the _______ .

**G** The fish is next to the _______ .

**H** The soap is below the _______ .

**I** The fish is _______ the salt and _______ the soap.

**J** The tea bags are above the _______ and next to the _______ .

▶ Copy and complete these sentences about this block of cubes.

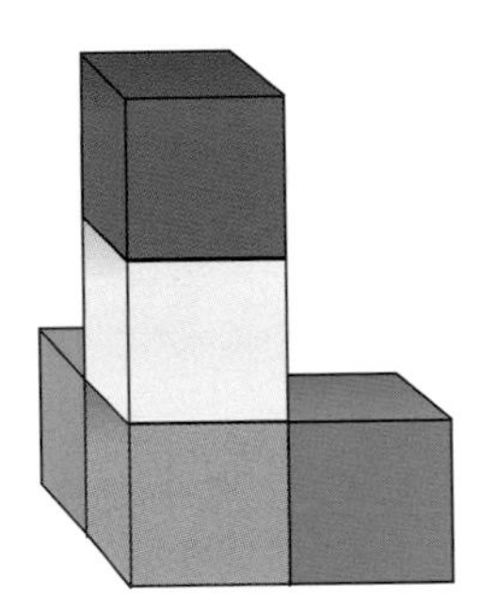

**K** The purple cube is _______ the white cube.

**L** The orange cube is _______ the grey cube.

**M** The pink cube is _______ the grey cube.

**N** The white cube is above the _______ cube.

**O** The grey cube is in front of the _______ cube and below the _______ cube.

▶ Write some sentences of your own about this block of cubes.

Choose from these words:

behind   next to   below
above   in front of

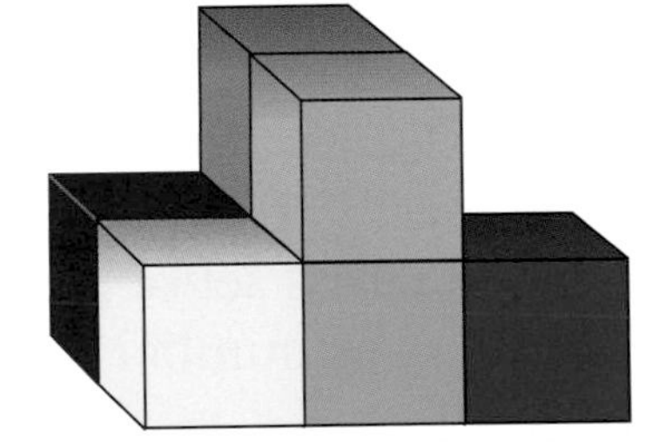

## Challenge

**1** Can you make this block of cubes?

The yellow cube is next to the red cube.
The red cube is in front of the green cube.
The white cube is above the red cube.

**2** Write instructions like these for a friend.

Can they make *your* block of cubes?

# Ordering numbers 1

## Using number lines

These numbers are jumbled up.
twenty-one eighteen twenty-five sixteen twenty
They have been marked on this number line.

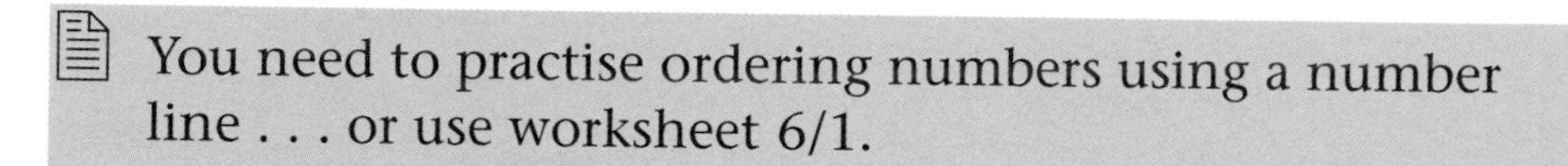


The numbers are **in order** on the number line.

**smallest** **largest**

16 18 20 21 25

You need to practise ordering numbers using a number line . . . or use worksheet 6/1.

► Write each set of numbers in order, from **smallest** to **largest**. Use your number line to help you.

| | | | | | |
|---|---|---|---|---|---|
| **A** | 2 | 7 | 3 | 5 | 9 |
| **B** | 1 | 10 | 6 | 0 | 8 |
| **C** | 8 | 12 | 14 | 5 | 7 |
| **D** | 25 | 7 | 33 | 18 | 45 |
| **E** | 42 | 63 | 3 | 24 | 44 |
| **F** | 17 | 49 | 50 | 27 | 30 |
| **G** | 17 | 21 | 18 | 20 | 23 |
| **H** | 52 | 58 | 50 | 53 | 59 |
| **I** | 43 | 39 | 32 | 30 | 40 |
| **J** | 82 | 20 | 28 | 88 | 80 |

▶ Write each set of numbers in order, from **largest** to **smallest**.

| | | | | | |
|---|---|---|---|---|---|
| **K** | 14 | 2 | 7 | 16 | 10 |
| **L** | 26 | 72 | 11 | 83 | 2 |
| **M** | 75 | 62 | 57 | 67 | 53 |
| **N** | 20 | 50 | 30 | 70 | 40 |
| **O** | 33 | 32 | 23 | 30 | 20 |
| **P** | 46 | 54 | 64 | 56 | 65 |

## In order

Year 9 are doing science projects.

▶ 9B are collecting leaves in science.
Here is the number of leaves they collected.

| Tom | Jan | Mina | Sam | Vin |
|---|---|---|---|---|
| 32 | 68 | 43 | 9 | 81 |

**1** Who collected the **largest** number?

**2** Who collected the **smallest** number?

**3** Write the numbers **in order**. Start with the **smallest**.

▶ 9C are counting snails for homework.
Here is the number of snails they counted.

| Kay | Amin | Dan | Zora | Ben |
|---|---|---|---|---|
| 25 | 10 | 17 | 40 | 28 |

**4** Who has the **greatest** number?

**5** Who has the **least** number?

**6** Write the numbers **in order**. Start with the **least**.

► 9A are counting petals on flowers.
Here is the number of petals they counted.

| Tony | Maya | Jim | Sue | Ros |
|---|---|---|---|---|
| 32 | 68 | 43 | 9 | 81 |

**7** Who has the **lowest** number?

**8** Who has the **highest** number?

**9** Write the numbers **in order**. Start with the **highest**.

► Here are the ages of their science teachers.

| Mrs Carter | Mr Ahmed | Ms Hall | Dr Smith | Mr Woods |
|---|---|---|---|---|
| 28 | 43 | 30 | 59 | 35 |

**10** Who is the **oldest**?

**11** Who is the **youngest**?

**12** Write the ages **in order**. Start with the **youngest**.

## Challenge

lowest youngest highest greatest
smallest least biggest
largest widest oldest shortest

Some of these words belong in pairs.
Find some more word pairs like these.

youngest ↔ oldest lowest ↔ highest

# Making hundreds

## Place value

There are ten blocks of ten cubes.
This makes **one hundred**.

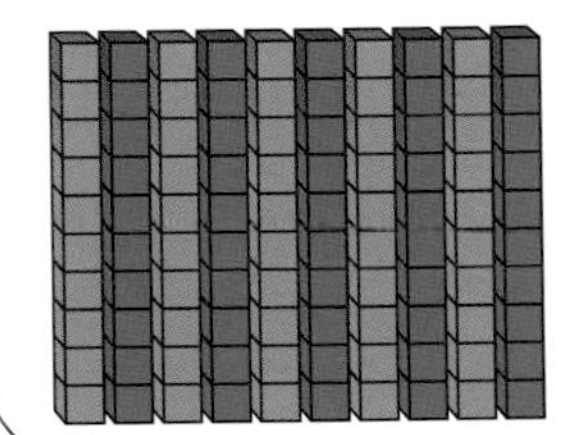

=

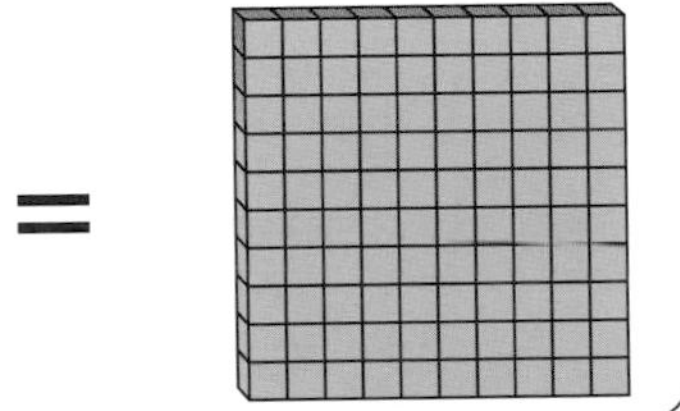

There are **327** cubes altogether in this picture.

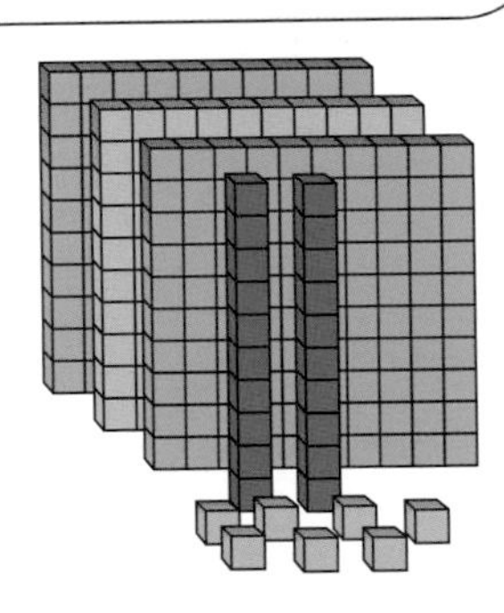

| | |
|---|---|
| 100 | one hundred |
| 200 | two hundred |
| 300 | three hundred |
| 400 | four hundred |
| 500 | five hundred |
| 600 | six hundred |
| 700 | seven hundred |
| 800 | eight hundred |
| 900 | nine hundred |
| 1000 | one thousand |

**327** means **300, 20 and 7** **3 hundreds, 2 tens** and **7 units**

You need to practise using hundreds, tens and units . . . or use worksheet 6/2.

▶ Copy and complete these sentences.

**A** **294** has ______ hundreds, ______ tens and ______ units.

**B** **317** has ______ hundreds, ______ tens and ______ units.

**C** ______ has **5** hundreds, **8** tens and **0** units.

**D** **905** has ______ hundreds, ______ tens and ______ units.

**E** ______ has **8** hundreds, **6** tens and **3** units.

# Unit 7

## Odd and even

► Copy and complete:

Odd numbers are

| 1 | 3 | 5 | 7 | | 11 | | | 17 | | |
|---|---|---|---|---|---|---|---|---|---|---|

Even numbers are

| 2 | 4 | 6 | 8 | | 12 | | | | 20 |
|---|---|---|---|---|---|---|---|---|---|

► Copy the table.
Put these numbers in the right place in the table.

19 5 17
12 9 16
8 4 0
3 18 1
10 13 2

| odd | even |
|---|---|
| | |

► Copy and complete.

Odd numbers go up in twos.

They end in 1 or 3 or ______ or ______ or ______ .

Even numbers go up in ______ .

They end in 0 or 2 or ______ or ______ or ______ .

**A** 20, 22, 24, ______, 28 are all even numbers

**B** 35, 37, 39, ______, 43 are all ______________ numbers

**C** 14, 16, 18, ______, 22 are all ______________ numbers

**D** 57, 59, 61, ______, ______ are all ______________ numbers

**E** 44, 46, 48, ______, ______ are all ______________ numbers

**F** 73, 75, 77, ______, ______ are all ______________ numbers

► For each number, write down **odd** or **even**.

**G** 6 **J** 38 **M** 47 **P** 93

**H** 17 **K** 45 **N** 54 **Q** 80

**I** 24 **L** 31 **O** 76 **R** 59

## Challenge

Can you find these numbers?

The next odd number after 25.

The next even number after 46.

Make up some problems like these yourself.

All the even numbers bigger than 55 and smaller than 60.

All the odd numbers between 70 and 80.

# Writing numbers 2

► Copy and complete this table.

| words | figures |
|---|---|
| forty-eight | |
| | 23 |
| thirteen | |
| | 81 |
| ninety-four | |
| | 12 |
| sixty-two | |
| | 75 |

# Change 2

## Change from 50p

One way to make 50p is like this.

How much change do you get from 50p if you spend 28p?

spend

28p

change

22p

You get 22p change.

▶ Work out how much change you get from 50p if you spend these amounts.
Use coins to help you.

| | | | |
|---|---|---|---|
| **A** | 16p | **F** | 7p |
| **B** | 39p | **G** | 48p |
| **C** | 42p | **H** | 33p |
| **D** | 11p | **I** | 24p |
| **E** | 26p | **J** | 15p |

## Change from £1

One way to make £1 is like this.

How much change do you get from £1 if you spend 72p?

spend

72p

change

28p

You get 28p change.

► Work out how much change you get from £1 if you spend these amounts.
Use coins to help you.

**K** 45p

**L** 77p

**M** 23p

**N** 64p

**O** 19p

**P** 56p

**Q** 82p

**R** 31p

# Subtracting 2

*you may use the 100 number line*

Kate is counting back in threes.

She starts at 20.

▶ Copy and complete these counting patterns.

**A** **15 14 13 12 11** _ _ _

**B** **88 86 84 82 80** _ _ _

**C** **90 85 80 75 70** _ _ _

**D** **32 28 24 20 16** _ _ _

**E** **60 57 54 51 48** _ _ _

▶ Make these counting patterns.

**F** Start at **30** . . . go down in **threes** . . . finish at **9**

**G** Start at 47 . . . go down in fours . . . finish at 27

**H** Start at 89 . . . go down in twos . . . finish at 71

**I** Start at 32 . . . go down in fives . . . finish at 7

**J** Start at 65 . . . go down in fours . . . finish at 49

# Five times table

| |
|---|
| $1 \times 5 = 5$ |
| $2 \times 5 = 10$ |
| $3 \times 5 = 15$ |
| $4 \times 5 = 20$ |
| $5 \times 5 = 25$ |
| $6 \times 5 = 30$ |
| $7 \times 5 = 35$ |
| $8 \times 5 = 40$ |
| $9 \times 5 = 45$ |
| $10 \times 5 = 50$ |

▶ Find the answers.

**A** $3 \times 5 =$

**B** $8 \times 5 =$

**C** $2 \times 5 =$

**D** $6 \times 5 =$

**E** $1 \times 5 =$

**F** $10 \times 5 =$

**G** $5 \times 5 =$

**H** ❈ $\times 5 = 15$

**I** ✤ $\times 5 = 20$

**J** ❁ $\times 5 = 5$

**K** ✱ $\times 5 = 45$

**L** ❈ $\times 5 = 30$

**M** ❄ $\times 5 = 50$

**N** ❈ $\times 5 = 35$

**O** ❄ $\times 5 = 40$

**P** seven times five

**Q** twice five

**R** six lots of five

**S** one times five

**T** double five

**U** nine multiplied by five

**V** ___ multiplied by five = twenty

## Challenge

There is a pattern in the times table.
Can you explain what the pattern is in words?
Tell your teacher.

# Is it half?

*you need squared paper*

**1** Which pictures show half of the square?

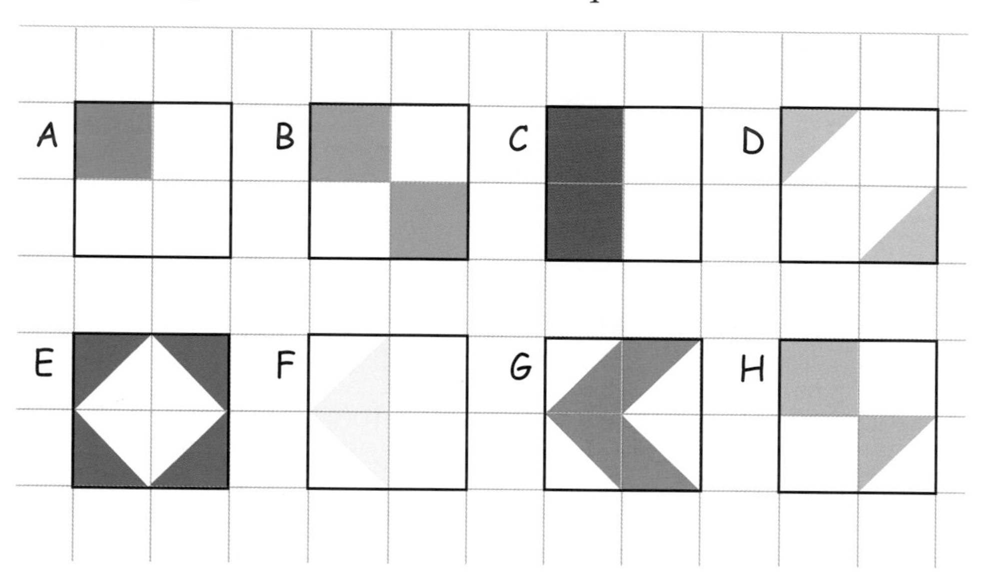


**2** Draw some more squares of your own.
Find different ways of shading exactly half.

**3** Display your work.
You could sort your patterns into different types.

# Adding 4

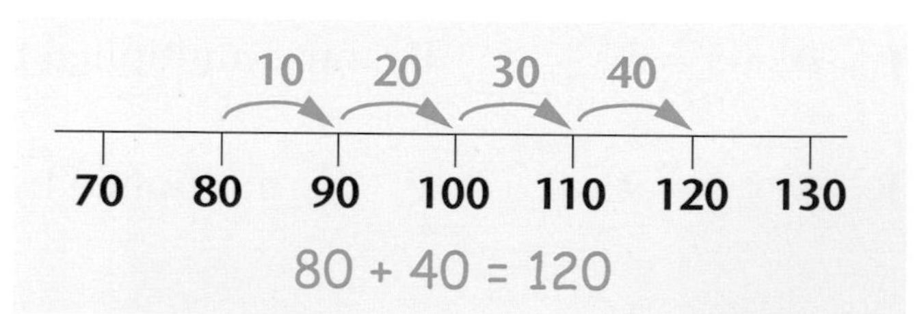


▶ Work these out without a calculator.

| | | | |
|---|---|---|---|
| **1** | 110 + 20 | **6** | 30 + 20 + 40 |
| **2** | 90 + 50 | **7** | 10 + 20 + 10 |
| **3** | 170 + 20 | **8** | 60 + 70 |
| **4** | 20 + 50 | **9** | 180 + 10 |
| **5** | 40 + 60 | **10** | 90 + 90 |

# Borrowing

23 subtract 7

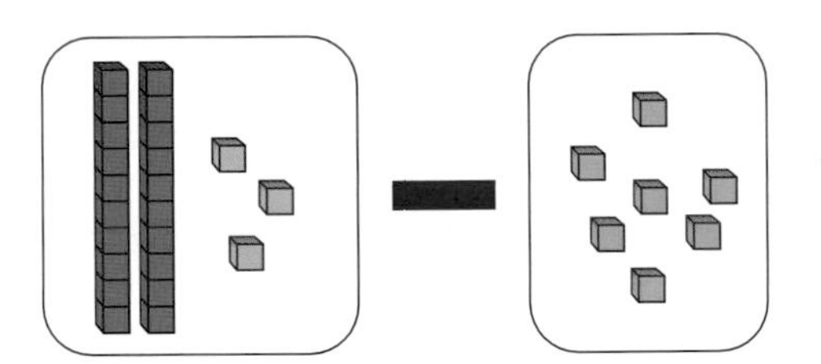

This sum cannot be done without breaking up a tens block

**Step 1**

20 and 3 subtract 7

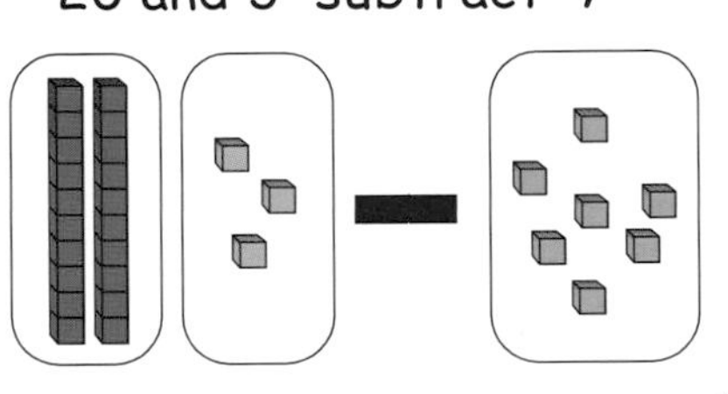

**Step 2**

**borrow one of the tens put it with the units**

10 and 13 subtract 7

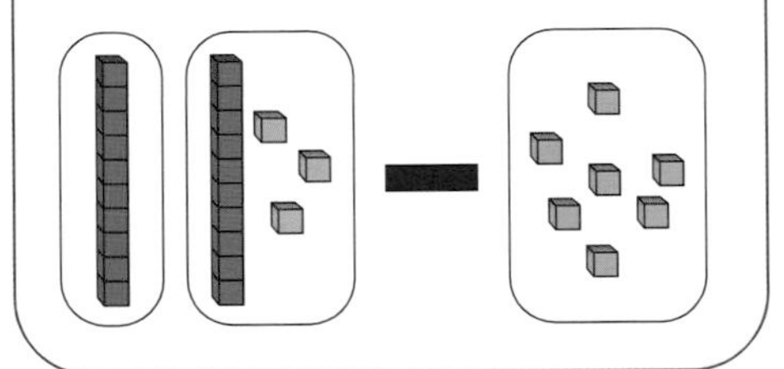

**Step 3**

**split up the ten to make 13 units**

10 and 13 subtract 7

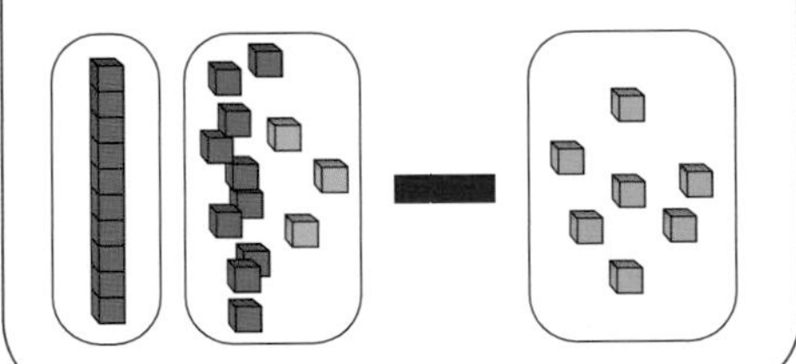

**Step 4**

**cross off the 7**

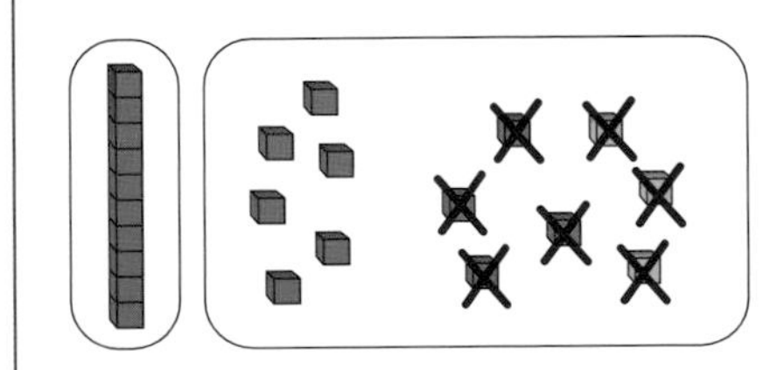

leaves 10 and 6 = 16

 You need to practise borrowing using cubes . . . or use worksheet 7/1.

# Unit 8

## Using tables 1

This table shows the number of sides for different shapes.

| Shape | Number of sides |
|---|---|
| parallelogram | 4 |
| pentagon | 5 |
| hexagon | 6 |
| octagon | 8 |
| decagon | 10 |

A pentagon has 5 sides.

▶ Answer these questions using the tables.

**1** Look at the table above.
**(a)** Which shape has 8 sides?
**(b)** How many sides does a parallelogram have?
**(c)** Which shape has 10 sides?
**(d)** How many sides does a hexagon have?

**2** This table shows the currency used in different countries.

| Country | Currency |
|---|---|
| Canada | Dollars |
| India | Rupees |
| Japan | Yen |
| Russia | Roubles |
| Turkey | Lira |

**(a)** Which currency is used in India?

**(b)** Where are Lira used?

**(c)** Which country uses Dollars?

**(d)** Which currency is used in Russia?

**3** This table shows Bank Holiday dates for 2001.

| Bank Holiday dates 2001 | |
|---|---|
| New Year | 1 January |
| Good Friday | 13 April |
| Easter Monday | 16 April |
| May Day | 7 May |
| Spring Bank Holiday | 28 May |
| Summer Bank Holiday | 27 August |
| Christmas Day | 25 December |
| Boxing Day | 26 December |

**(a)** What date is the Spring Bank Holiday?

**(b)** What holiday is on 27 August?

**(c)** What date is Boxing Day?

**(d)** What holiday is on 16 April?

**(e)** When is the Spring Bank Holiday?

# Solid shapes 2

*you may need the solids kit*

cuboid cylinder face curved face edge
corner vertex flat rounded pointed

This is a **cuboid**.

A **face** is one **surface** of the solid.
A **vertex** is a **corner** of the solid.

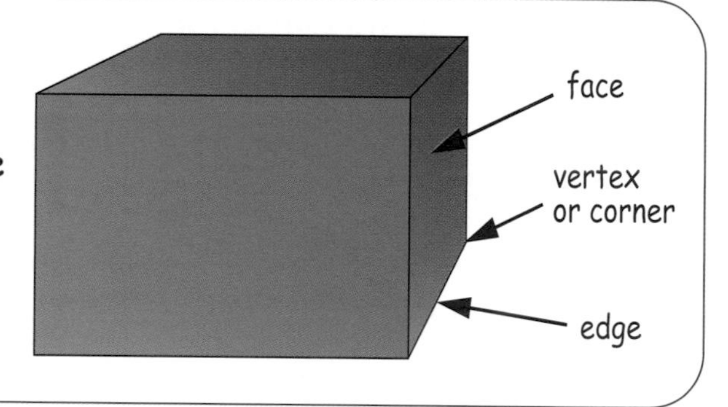


▶ Answer these questions about the cuboid.
Remember to think about the parts you can't see.

**1** How many faces does the cuboid have?

**2** How many corners does the cuboid have?

**3** How many edges does the cuboid have?

**4** Draw the solid. Mark each corner with X.
Can you see them all?

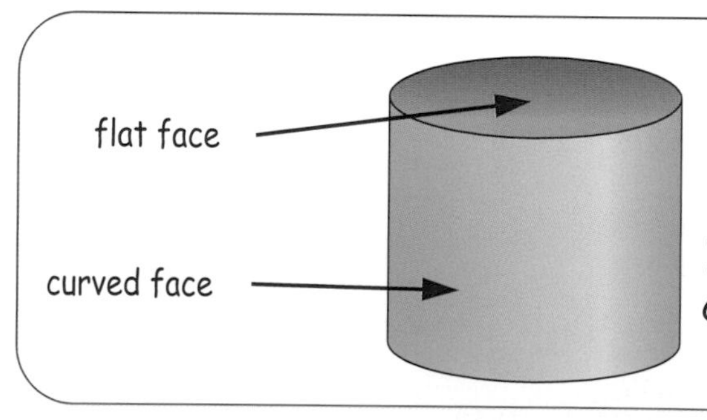


This is a **cylinder**.

It has a curved face and flat faces.
If you put it down on its curved face it would **roll**.
It has no corners.

▶ Answer these questions about the cylinder.

**5** How many flat faces does it have?

**6** How many curved faces does it have?

▶ Copy these solids and complete the sentences.

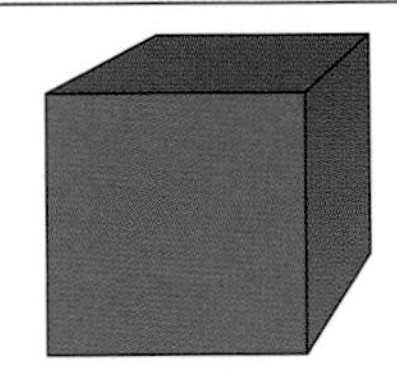

This is a cube.
It has **12** edges.
It has ___ faces.
It has ___ corners.

This is a cylinder.
It has ___ flat faces.
It has ___ curved faces.
It has ___ corners.

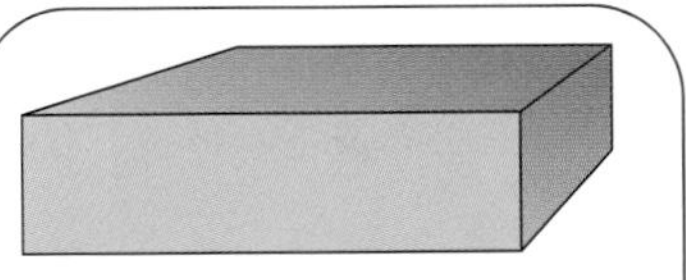

This is a cuboid.
It has ___ edges.
It has ___ faces.
It has **8** corners.

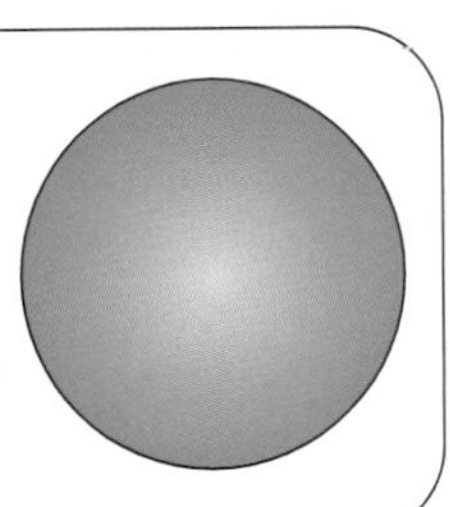

This is a sphere.
It has **no** edges.
It has ___ flat faces.
It has ___ curved faces.
It has ___ corners.

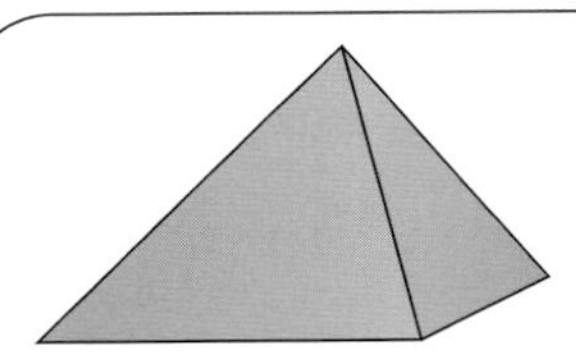

This is a pyramid.
It has ___ faces.
It has ___ edges.
It has ___ corners.

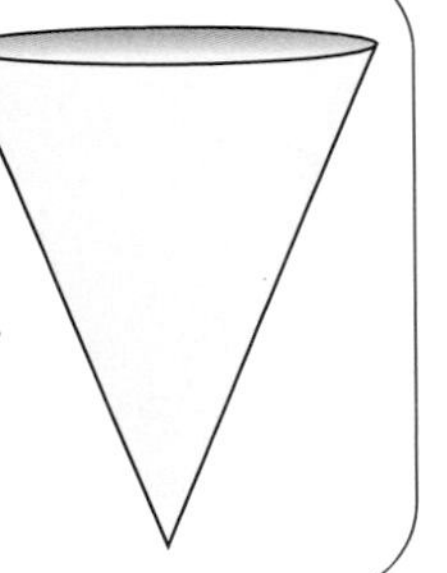

This is a cone.
It has ___ flat faces.
It has ___ curved faces.
It has ___ vertex.

## Challenge

Find as many different solids as you can.
How many faces do they have?
How many corners do they have?
Can you draw them?

You need to practise describing solid shapes . . . or use worksheet 8/1.

# Reading scales 3

## Review

*Look back at page 46 if you cannot answer this question easily.*

▶ Write down the number each arrow points to.

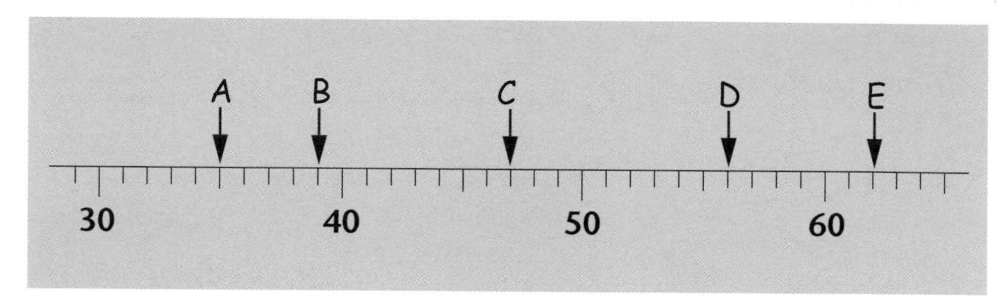


## Reading scales

These scales do **not** go up in **ones**.

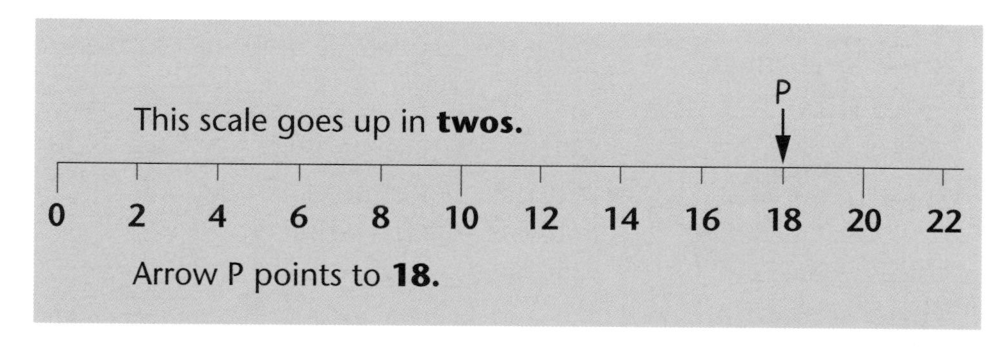


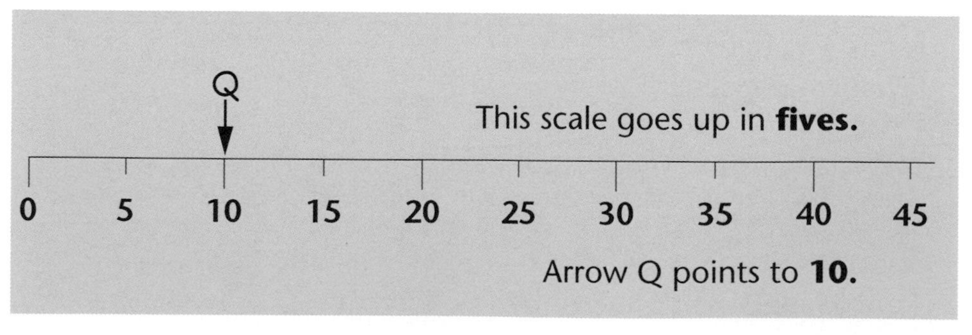

You need to practise numbering scales in 2s and 5s . . . or use worksheet 8/2, section A.

▶ Write down the number each arrow points to.

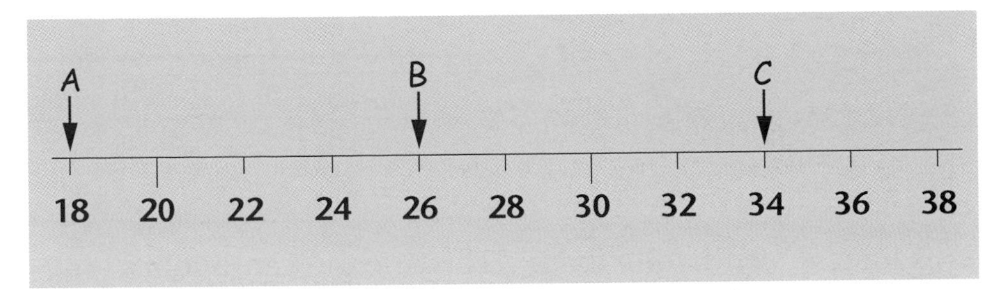


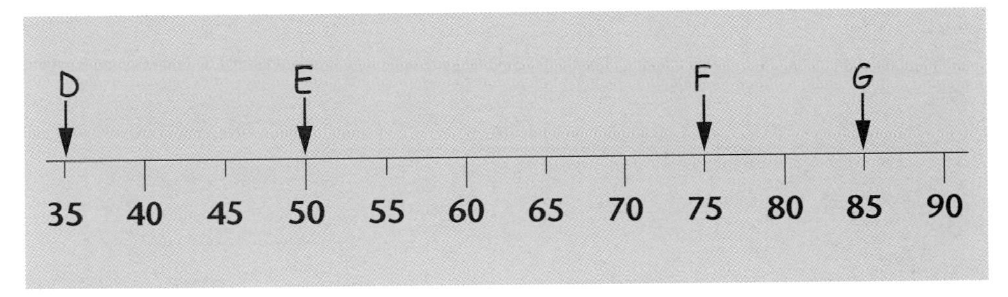


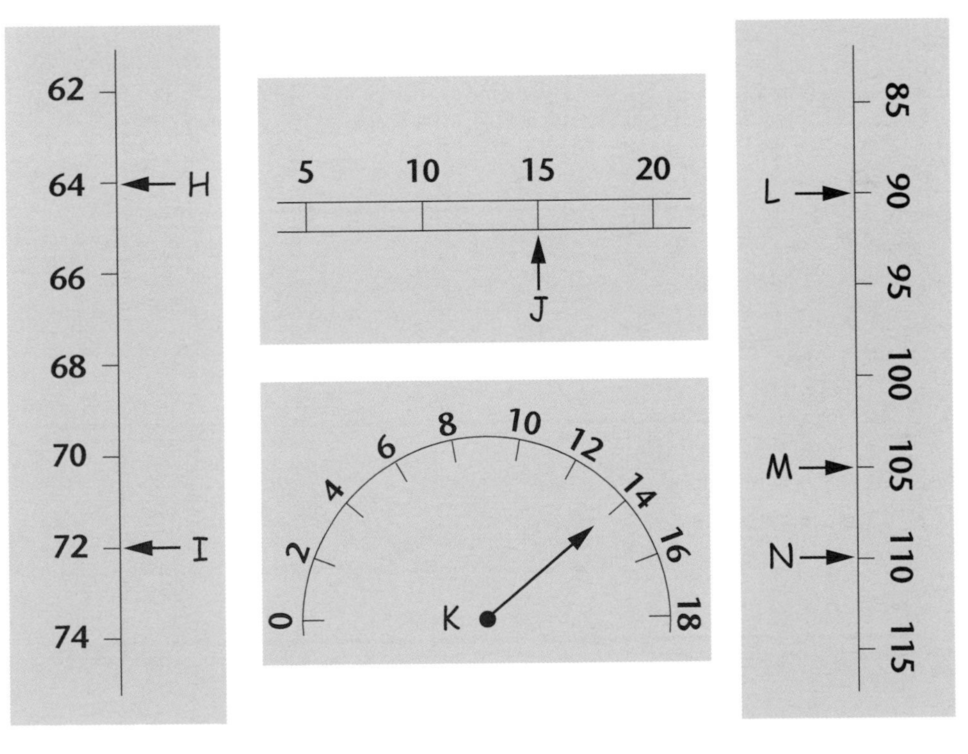


You need to practise marking arrows on these scales . . . or use worksheet 8/2, section B.

# Lines 1

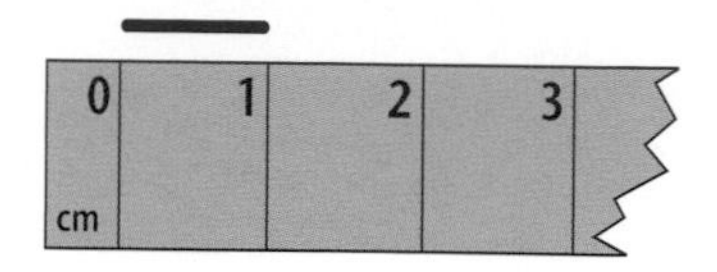


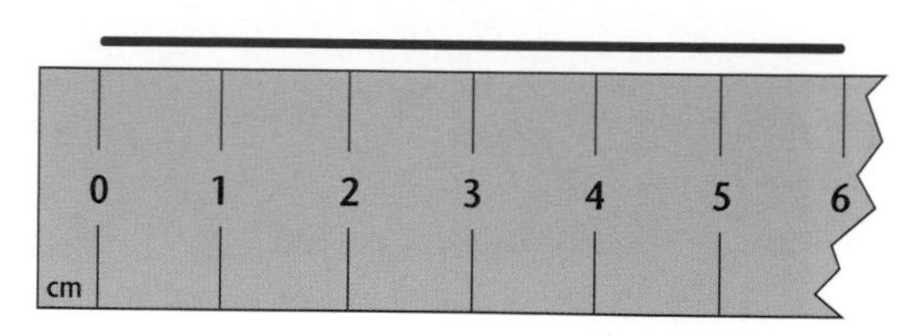


This line is 1 centimetre long. This line is 6 centimetres long.

▶ How many centimetres long is each line?

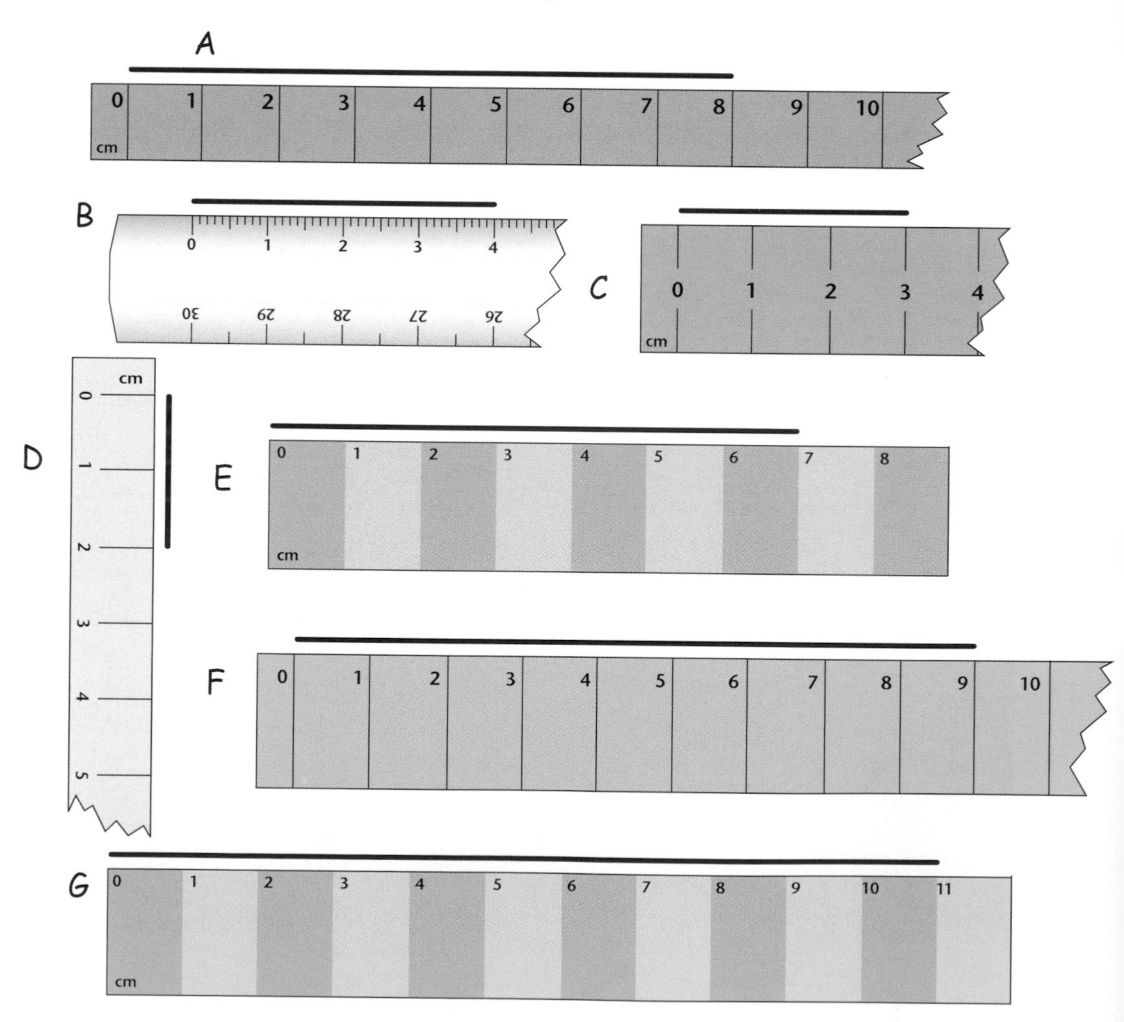


You need to practise measuring lines . . . or use worksheet 8/3.

## Estimating length

———————————— This line is 5 centimetres long.

**1** Look at the lines on this page.
Copy this table.
Complete it without measuring the lines.

| Shorter than 5 centimetres | Longer than 5 centimetres |
|---|---|
| | |

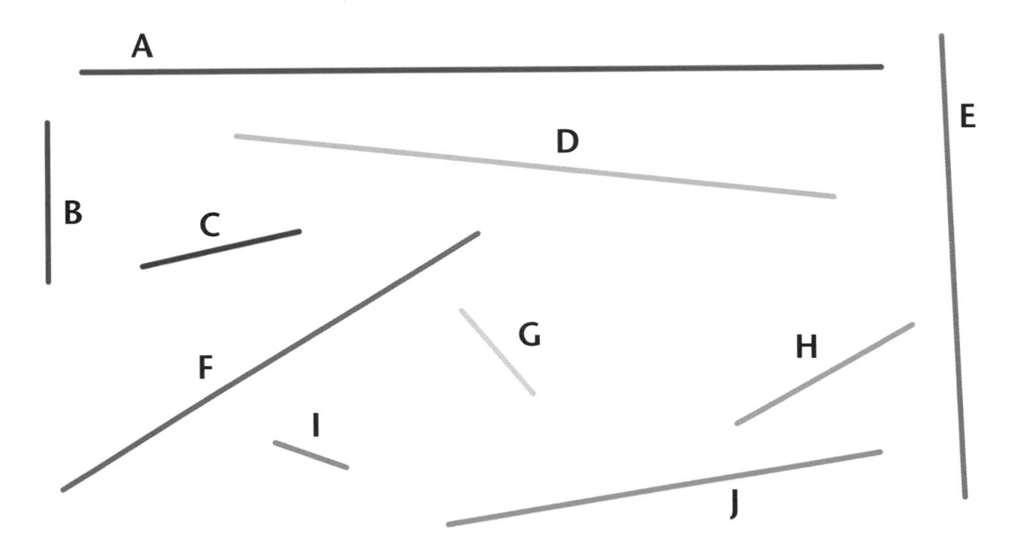


**2** Which line is about 10 centimetres long?

### Challenge

*you need the 50 centimetre tape*

Write down 5 things that are longer than 50 centimetres.
Write down 5 things that are shorter than 50 centimetres.

Now check with your tape.
Were you right?

Can you find something 50 centimetres long?
How close were you?

# Right angles 2

An angle is made where 2 lines meet.

A right angle is a special sort of angle.

All of the angles shown are right angles.
You can check with the corner of a piece of paper.

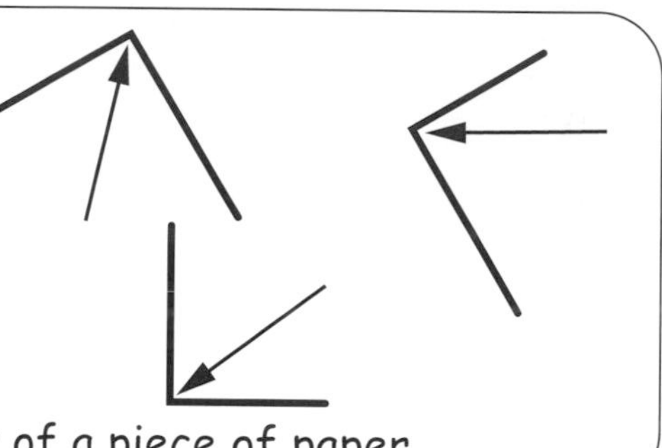

▶ Look at the angles below.

Are they **smaller than** a right angle, a **right angle**, or **bigger than** a right angle?

▶ Write smaller, right angle or bigger for each.

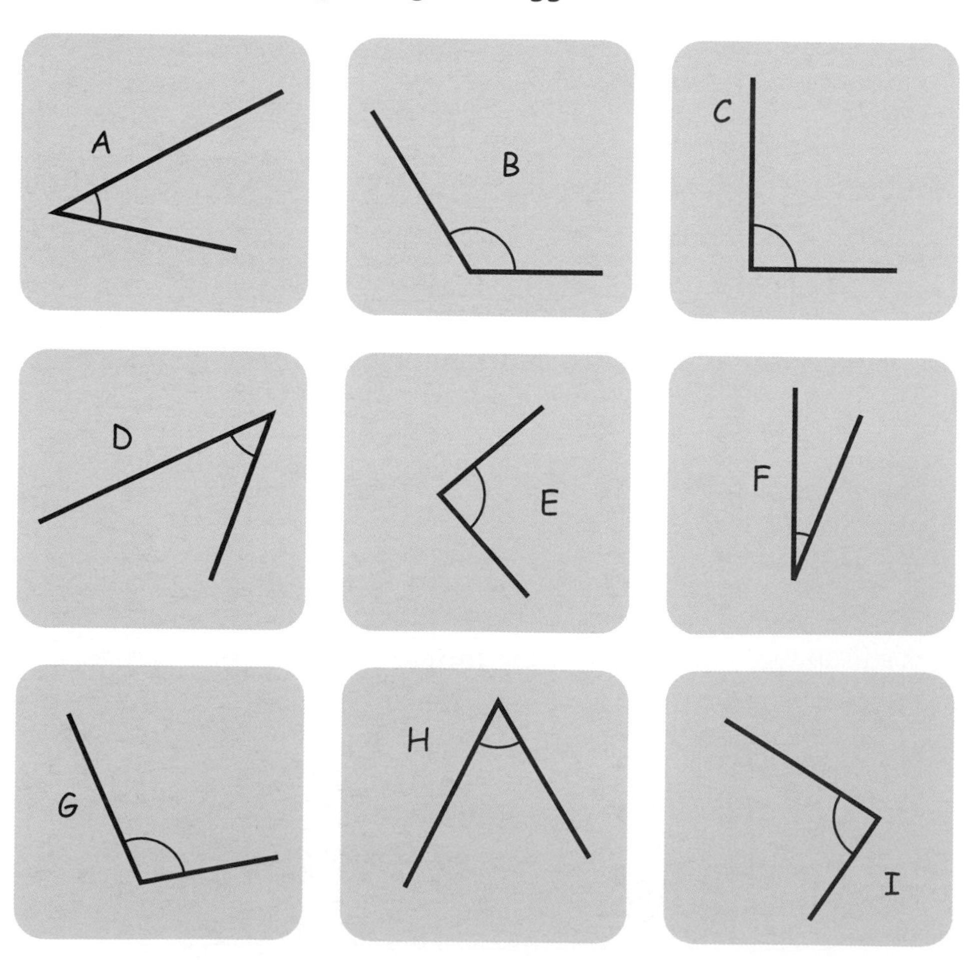

## Angles in shapes

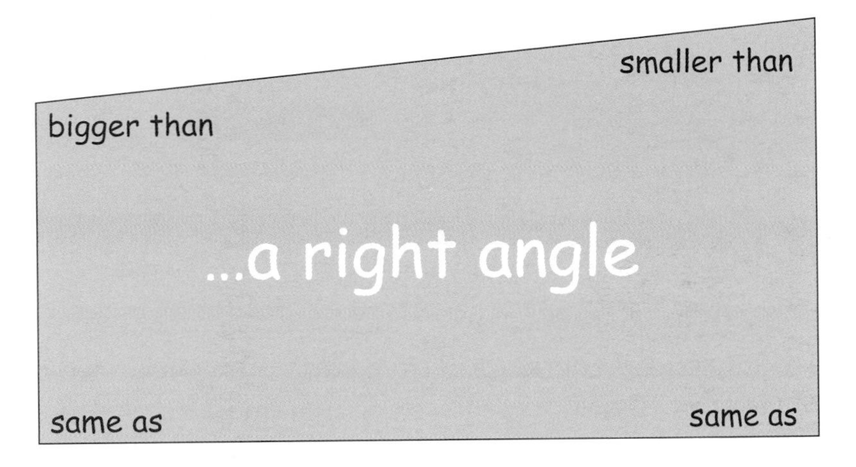


- Look at the marked angles in these shapes.

  Are they **smaller than** a right angle, **the same as** a right angle, or **bigger than** a right angle?

- Write smaller, right angle or bigger for each.

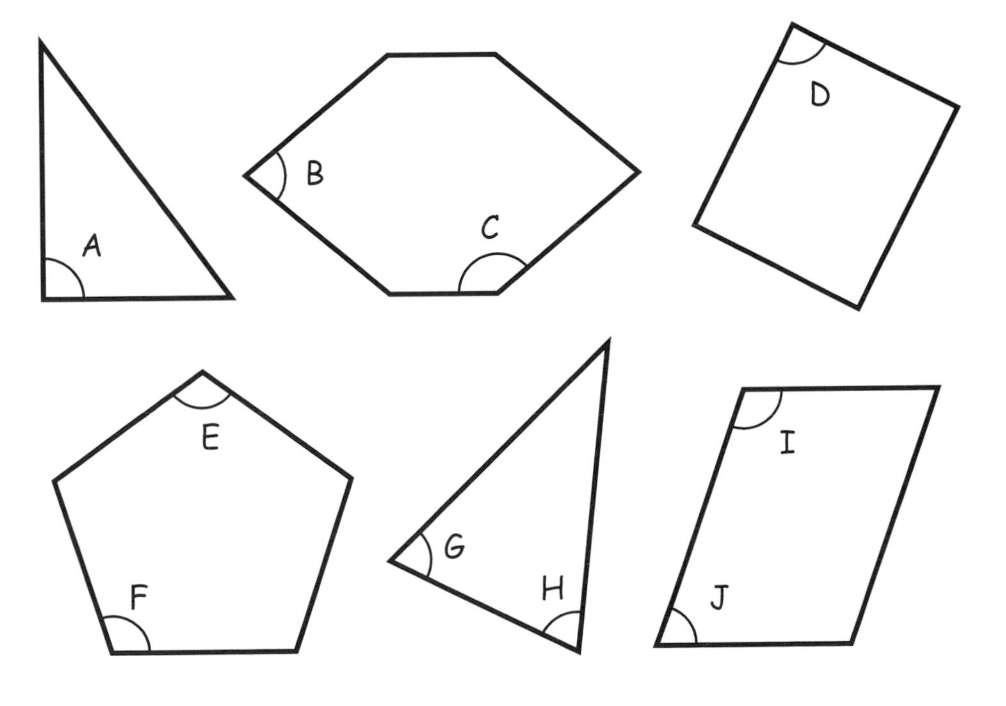

You need to practise describing angles in shapes . . . or use worksheet 8/4.

# Halves and quarters 1

Each of these shapes is split into **four equal parts**.
Each part is called **one quarter**.

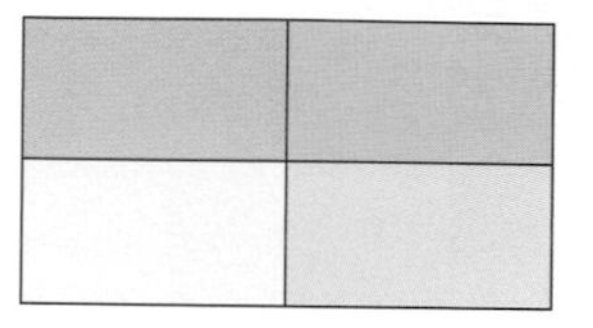 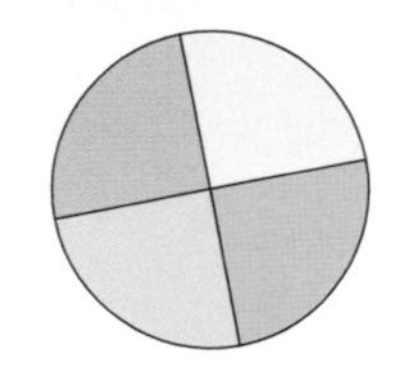 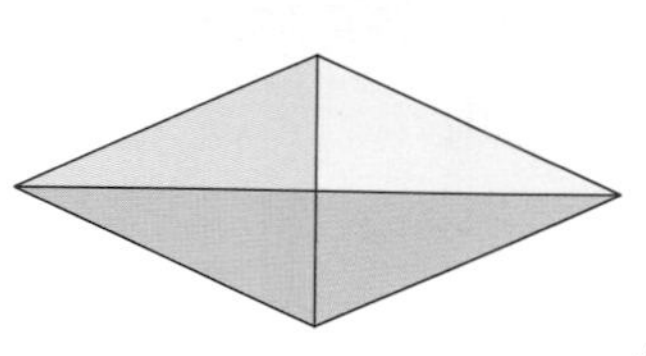

One part out of four is green.
One quarter of this shape is green.

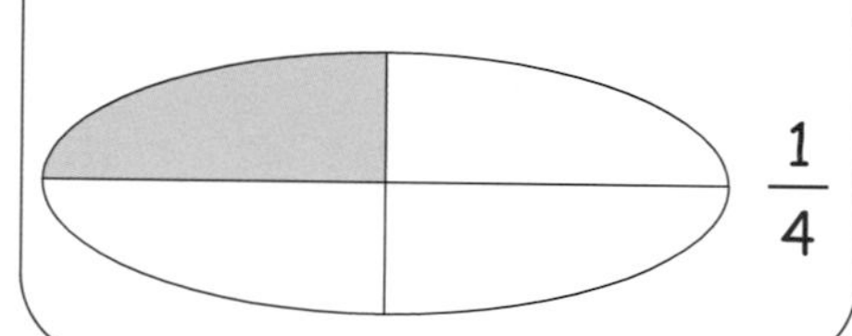

$\frac{1}{4}$

Three parts out of four are blue.
Three quarters of this shape are blue.

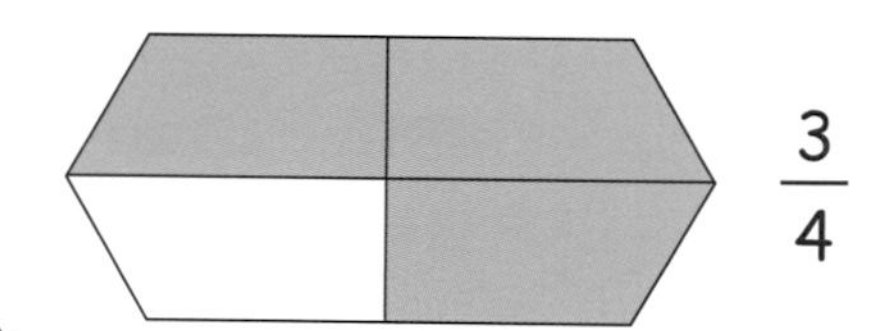

$\frac{3}{4}$

► Write down the fraction of each shape that has been coloured.
The first one has been done for you.

**A**

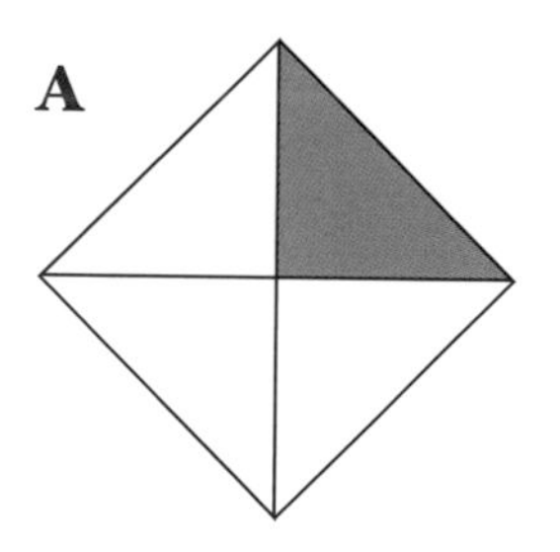

one quarter

**B**

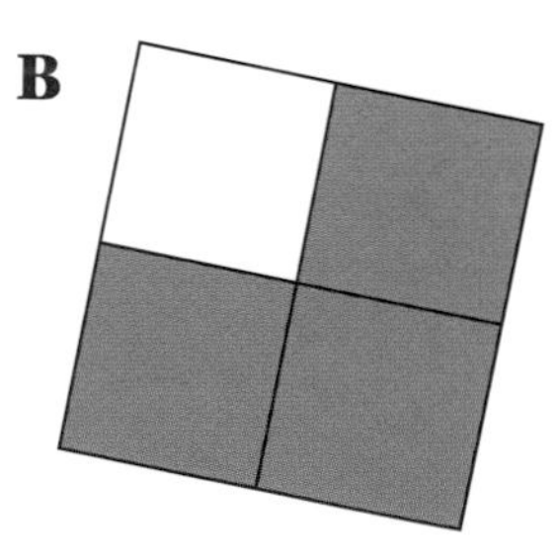

**C**

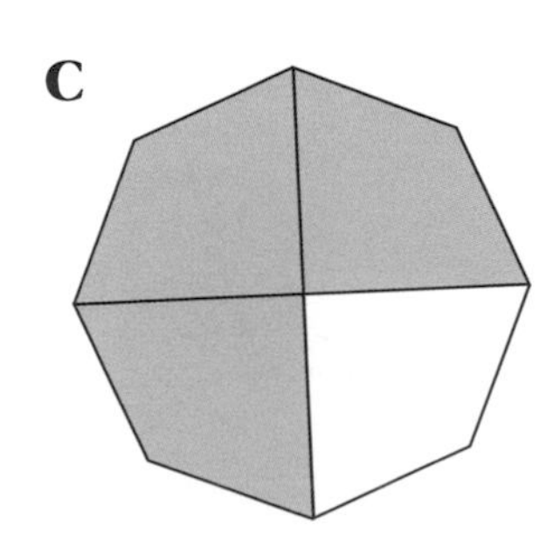

**D**

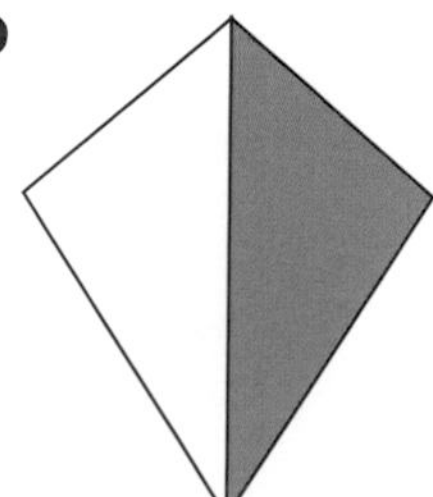

**E**

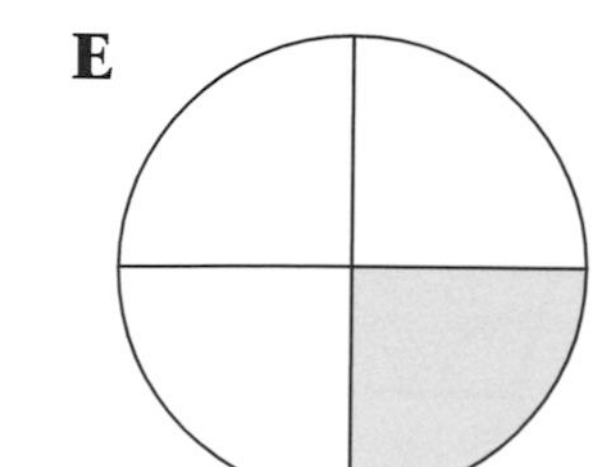

 You need to practise shading $\frac{1}{2}$, $\frac{1}{4}$ or $\frac{3}{4}$ of shapes . . . or use worksheet 8/5.

# Measuring metres

*you need a metre stick and a piece of chalk*

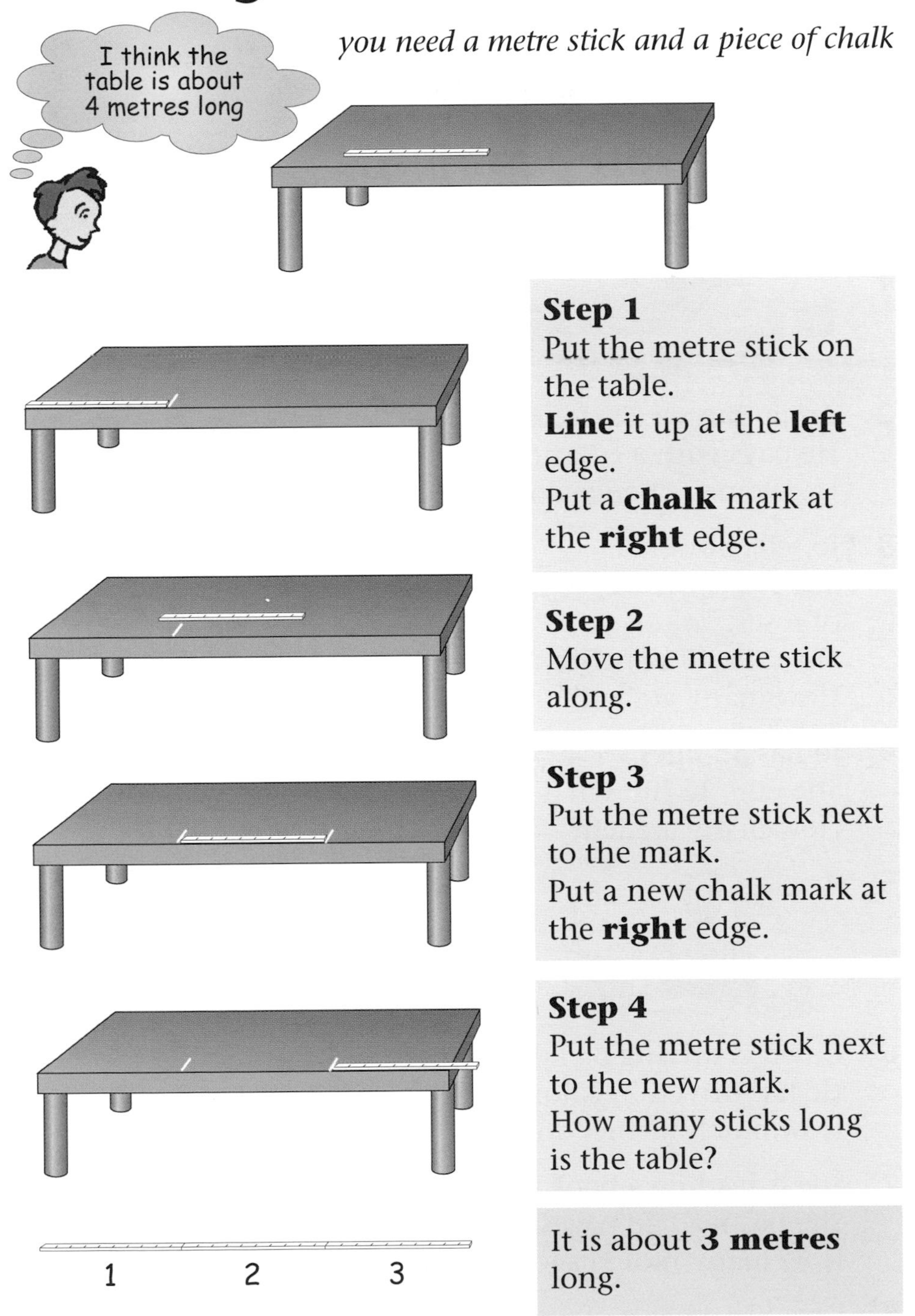


**Step 1**
Put the metre stick on the table.
**Line** it up at the **left** edge.
Put a **chalk** mark at the **right** edge.

**Step 2**
Move the metre stick along.

**Step 3**
Put the metre stick next to the mark.
Put a new chalk mark at the **right** edge.

**Step 4**
Put the metre stick next to the new mark.
How many sticks long is the table?

It is about **3 metres** long.

You need to practise estimating and measuring lengths in metres . . . or use worksheet 8/6.

# Unit 9

## Subtraction problems

### Work it out

**1** 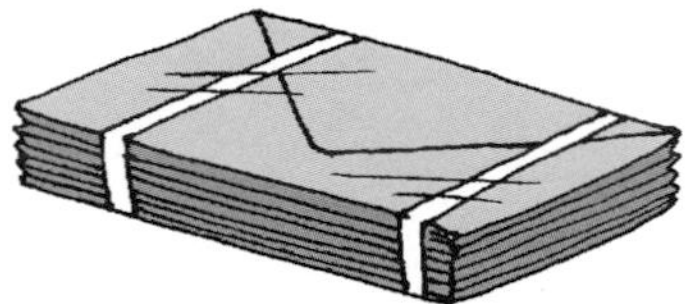

Leah opens a pack of 60 envelopes.
She uses 18 envelopes.
How many are left?

**2** Amjid buys a CD for £12.
He pays with a £20 note.
How much change does he get?

**3** Jo sells flowers.
She starts the day with 75 bunches of roses.
She sells 48 bunches.
How many are left?

**4** Jo has 56 lilies.
She puts 12 lilies in a bunch.
How many are left?

**5** 

Connor is cooking for 47 people.
He has made 36 meals.
How many more does he have to make?

**6** David is driving 65 kilometres to visit his mother.
He has driven 29 kilometres.
How much further does he have to go?

**7** Sue is reading a book with 92 pages.
She has read 48 pages so far.
How many more pages does she have to read?

**8** Faye has a piece of wood 85 centimetres long.
She cuts off 20 centimetres.
How long is the piece of wood now?

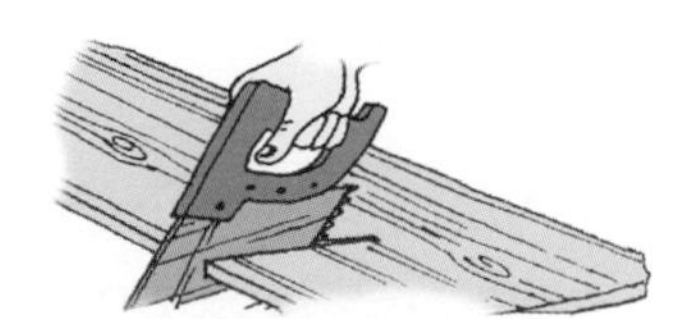

## Number problems

**A** What number do you add to 36 to make 80?

**B** Find the difference between 41 and 72.

**C** What number is 12 less than 68?

**D** What number is 57 less than 93?

**E** Find the difference between 18 and 81.

**F** The total of two numbers is 48.
One of the numbers is 26.
What is the other number?

**G** What number do you add to 29 to make 65?

**H** The total of two numbers is 81.
One of the numbers is 34.
What is the other number?

### Missing numbers

Can you find the missing numbers in these puzzles?

```
  46        29        27
+  *      +  *      + **
----      ----      ----
  52        37        55

  38        71        94
-  *      + **      - **
----      ----      ----
  29        97        68
```

# Bar charts 1

Lucy asked her friends to choose their favourite day out.

She displayed the results in this bar chart.

The bar chart shows that 4 people chose seaside as their favourite day out.

► Look at Lucy's bar chart.

**1** How many people chose London?

**2** How many people chose shopping?

**3** How many people did Lucy ask altogether?

**4** The **most popular** day out was theme park. It has the **biggest** bar in the chart.

The **least popular** has the **smallest** bar. Which day out was least popular?

▶ A café asked people if they preferred tea or coffee. This bar chart shows the results.

5 How many people chose tea?

6 How many people chose coffee?

7 How many people did not mind?

8 How many people were asked altogether?

▶ Danny asked his friends about their favourite type of film. This bar chart shows his results.

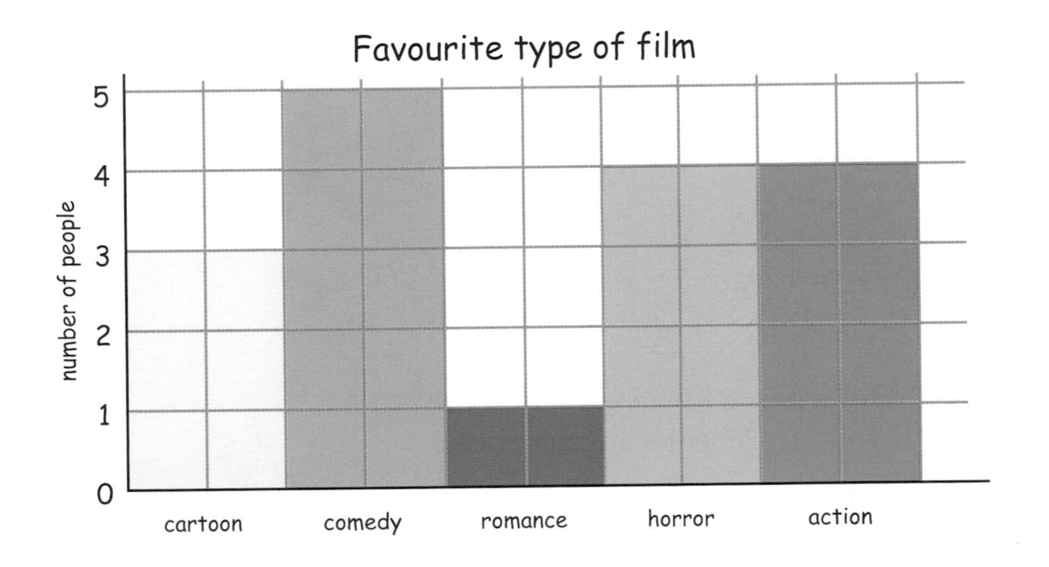


9 How many people chose horror?

10 Which type of film was most popular?

11 How many people were asked altogether?

# Ten times table

| |
|---|
| $1 \times 10 = 10$ |
| $2 \times 10 = 20$ |
| $3 \times 10 = 30$ |
| $4 \times 10 = 40$ |
| $5 \times 10 = 50$ |
| $6 \times 10 = 60$ |
| $7 \times 10 = 70$ |
| $8 \times 10 = 80$ |
| $9 \times 10 = 90$ |
| $10 \times 10 = 100$ |

▶ Work these out.

**A** $4 \times 10$

**B** $3 \times 10$

**C** $7 \times 10$

**D** $10 \times 10$

**E** $1 \times 10$

**F** $5 \times 10$

▶ Copy these and fill in the missing numbers.

**G** ✷ $\times 10 = 60$

**H** $2 \times$ ✷ $= 20$

**I** ✷ $\times 10 = 90$

**J** $8 \times$ ✷ $= 80$

▶ Work these out. Write down the sum you do each time.

**K** 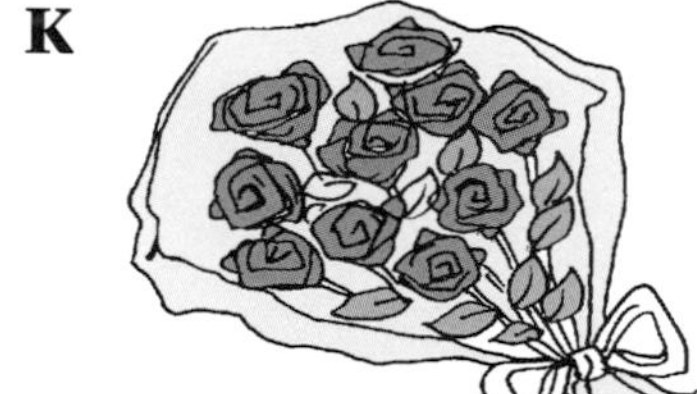

There are 10 roses in a bunch.
How many roses are there in 6 bunches?

**L** Biscuits come in packs of 10.
How many biscuits are there in 4 packs?

**M** Pens cost 10p each.
How much do 8 pens cost?

**N** A CD rack holds 10 CDs.
How many CDs do 3 racks hold?

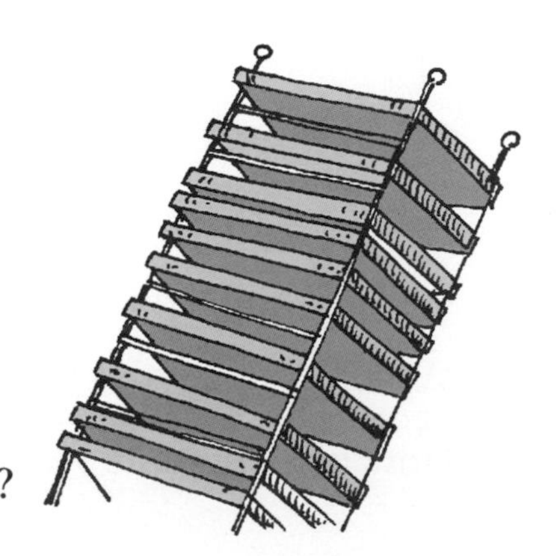

# Drawing shapes 3

Follow these steps to draw a pentagon without a grid.

**Step 1**
Put 5 dots on the paper.
A pentagon has
**5** sides and
**5** corners.

**Step 2**
Join the dots with a ruler.

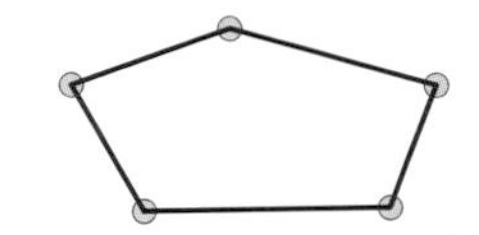

▶ Draw these shapes without a grid.

**A** triangle
**B** hexagon
**C** octagon
**D** rectangle
**E** square
**F** star

## Challenge

One way to draw a circle is to draw round a coin.
How many different ways can you find to draw a perfect circle?

# Writing numbers 3

▶ Copy and complete this table.

| words | figures |
|---|---|
| two hundred and thirty-eight | |
| | 600 |
| five hundred and seventy-three | |
| | 758 |
| one hundred and forty-six | |
| | 819 |
| four hundred and twenty | |
| | 302 |

# Mirrors 1

*you will need a mirror*

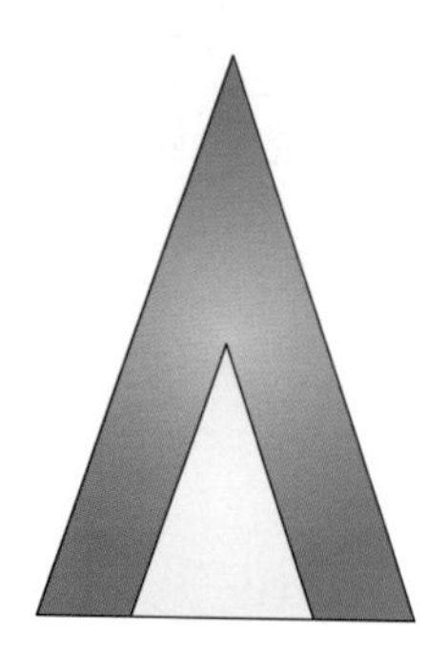

Put your mirror on top of this pattern. It looks the same when the mirror is on top.

It has **reflection symmetry**.

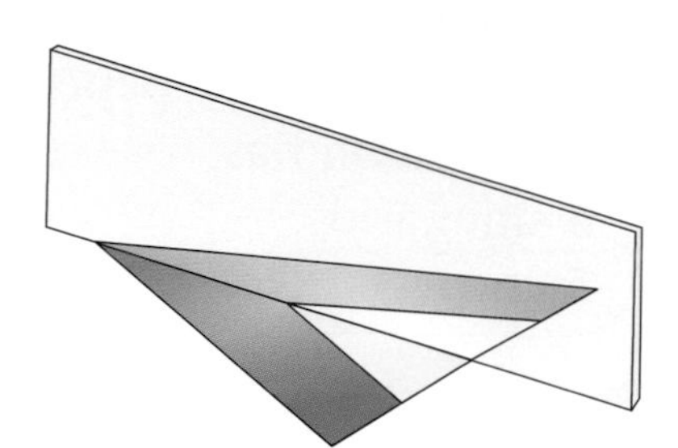

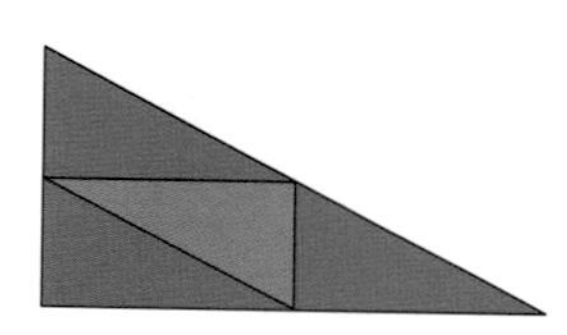

Put your mirror on top of this pattern. It looks different when the mirror is on top.

It does **not** have reflection symmetry.

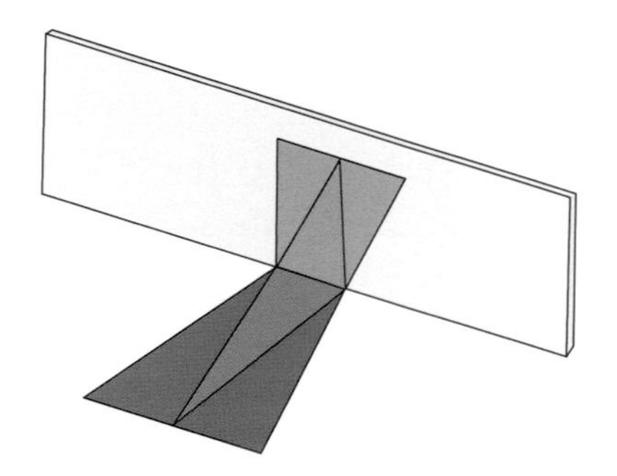

► Which of these patterns have reflection symmetry? Write yes or no.

**A**

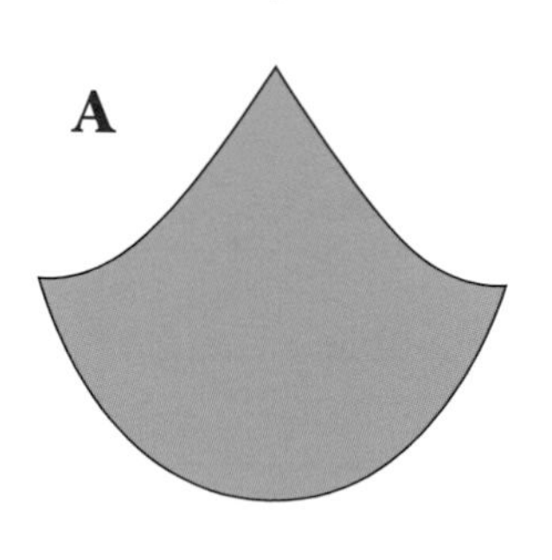

**B**

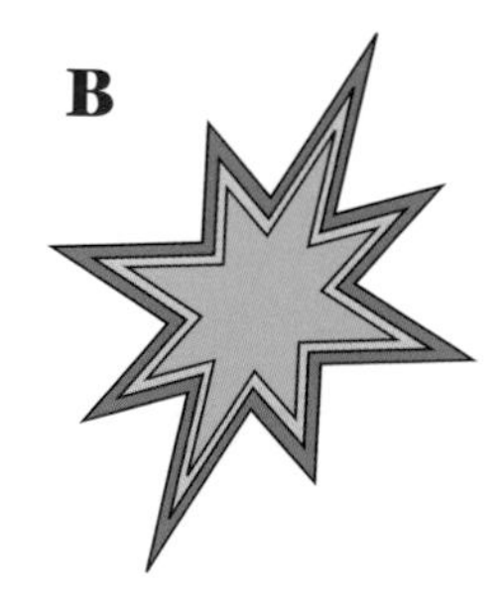

**C**

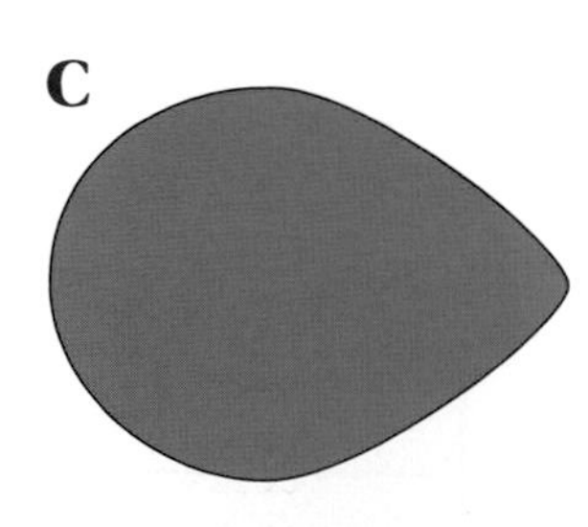

**D**

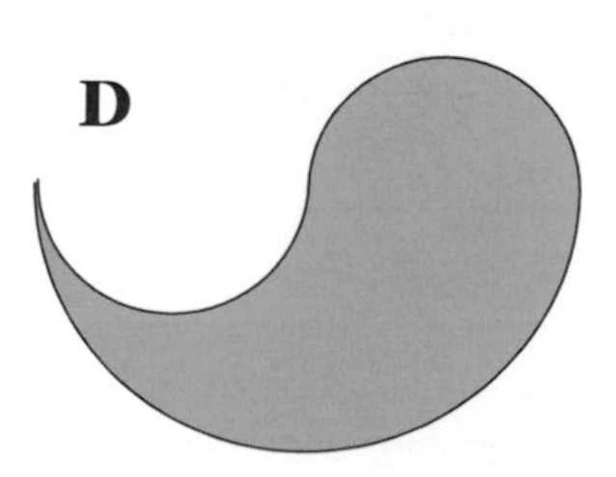

**E**

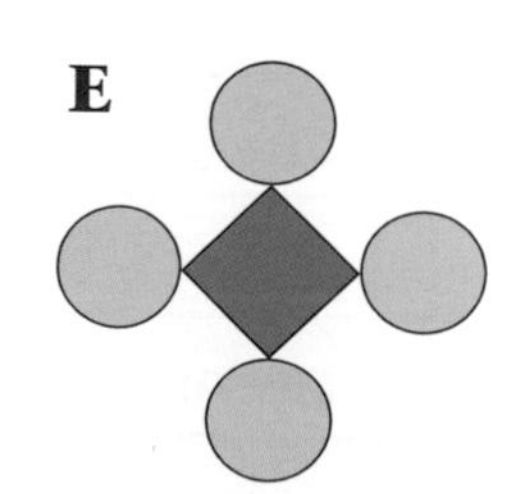

**F**

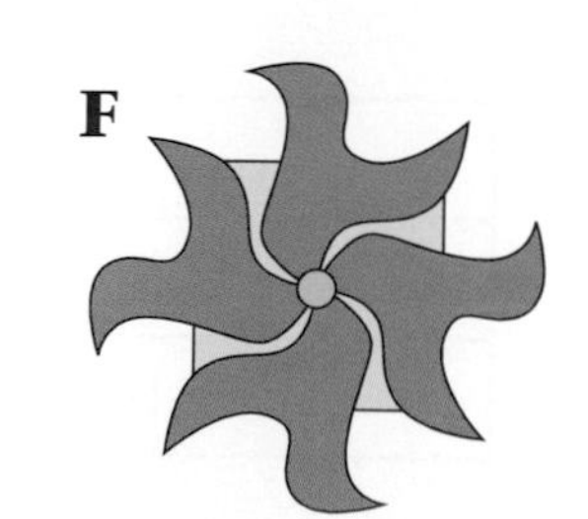

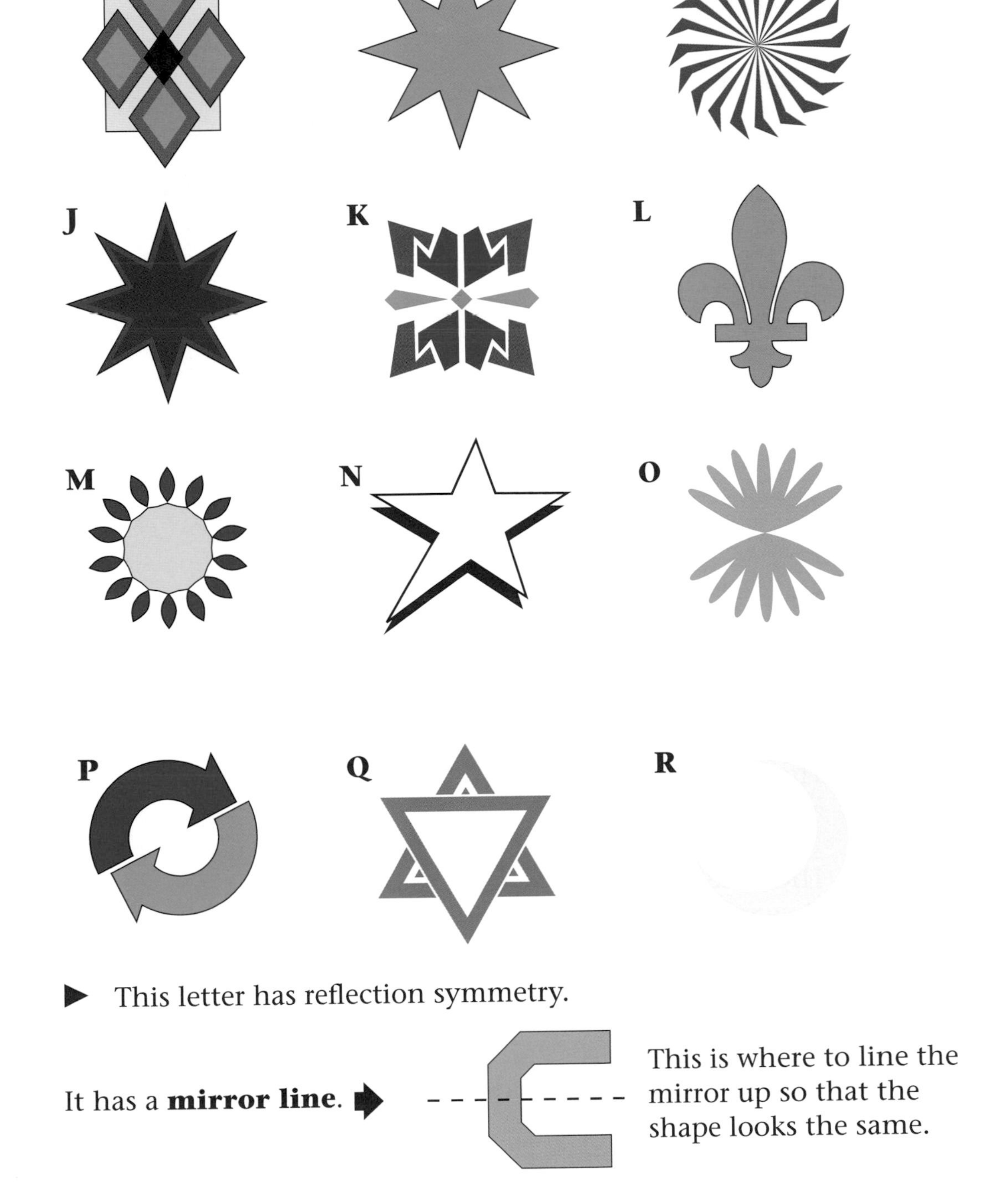


▶ This letter has reflection symmetry.

It has a **mirror line**. ➡

This is where to line the mirror up so that the shape looks the same.

You need to practise finding mirror lines on shapes . . . or use worksheet 9/1.

# Multiplying

**1** What is double six?

**2** There are 4 cakes in a box.
How many cakes are there in 3 boxes?

**3** There are 5 cards in a pack.
How many cards are there in 10 packs?

**4** What is twice ten?

**5** Work out 3 lots of 8.

**6** Work out 5 multiplied by 7.

**7** There are 3 seats in a rollercoaster car.
How many seats are there in 5 cars?

**8** How many people are there in 4 five-a-side football teams?

**9** Work out 4 times 9.

**10** There are 2 pens in a pack.
How many pens are there in 8 packs?

**11** There are 3 cartons of juice in a pack.
How many cartons are there in 6 packs?

**12** What is double 7?

**13** A cinema ticket costs £5.
How much do 6 tickets cost?

**14** What is four multiplied by seven?

**15** Work out 3 lots of 9.

3 times table

$1 \times 3 = 3$
$2 \times 3 = 6$
$3 \times 3 = 9$
$4 \times 3 = 12$
$5 \times 3 = 15$
$6 \times 3 = 18$
$7 \times 3 = 21$
$8 \times 3 = 24$
$9 \times 3 = 27$
$10 \times 3 = 30$

4 times table

$1 \times 4 = 4$
$2 \times 4 = 8$
$3 \times 4 = 12$
$4 \times 4 = 16$
$5 \times 4 = 20$
$6 \times 4 = 24$
$7 \times 4 = 28$
$8 \times 4 = 32$
$9 \times 4 = 36$
$10 \times 4 = 40$

# Unit 10

## Missing numbers 2

▶ There is a number missing from each □.
Copy and complete these sums.

| | | |
|---|---|---|
| **A** 10 + 5 = □ | **E** 14 + 2 = □ | **I** 11 + 1 = □ |
| **B** 12 + 6 = □ | **F** 11 + 9 = □ | **J** 13 + 5 = □ |
| **C** 17 + 3 = □ | **G** 16 + 2 = □ | **K** 15 + 4 = □ |
| **D** 8 + 12 = □ | **H** 4 + 13 = □ | **L** 7 + 10 = □ |

▶ Each shape in these sums stands for a different number.
Copy and complete them.

| | | |
|---|---|---|
| **M** 19 + 1 = ✱ | ✱ − 19 = 1 | ✱ − 1 = 19 |
| **N** 15 + 2 = ■ | ■ − 15 = 2 | ■ − 2 = 15 |
| **O** 10 + ● = 16 | 16 − ● = 10 | 16 − 10 = ● |
| **P** ◆ + 2 = 19 | 19 − ◆ = 2 | 19 − 2 = ◆ |
| **Q** ✤ + 5 = 17 | 17 − ✤ = 5 | 17 − 5 = ✤ |
| **R** ❊ + 7 = 18 | 18 − ❊ = 7 | 18 − 7 = ❊ |
| **S** 18 + 2 = ✱ | ✱ − 18 = 2 | 20 − 2 = ✪ |
| **T** 16 + 3 = ▼ | 19 − 16 = ◗ | 19 − ◗ = 16 |
| **U** ✦ + 6 = 20 | 20 − ▲ = 14 | 20 − ✦ = 6 |
| **V** 13 + 7 = ✱ | 20 − 13 = ✶ | 20 − ✶ = 13 |

# Temperatures 1

This thermometer is used to measure temperature in degrees Celsius.
The scale is marked every 10 degrees.
Each small division is equal to 1 degree Celsius.

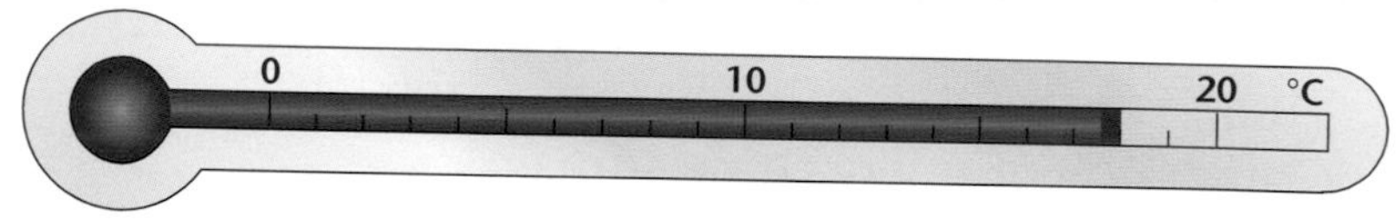


The thermometer shows 18 degrees Celsius.
We can write this as **18 °C**.

▶ Write down the temperature shown on each of these thermometers.

**A**

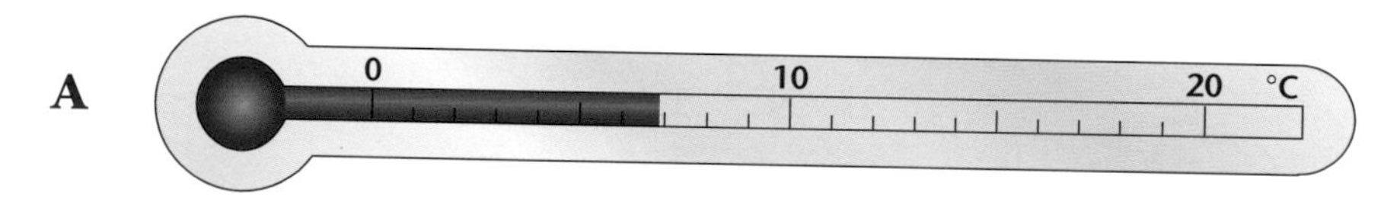


Write your answers like this:

A  The temperature is 7 °C.

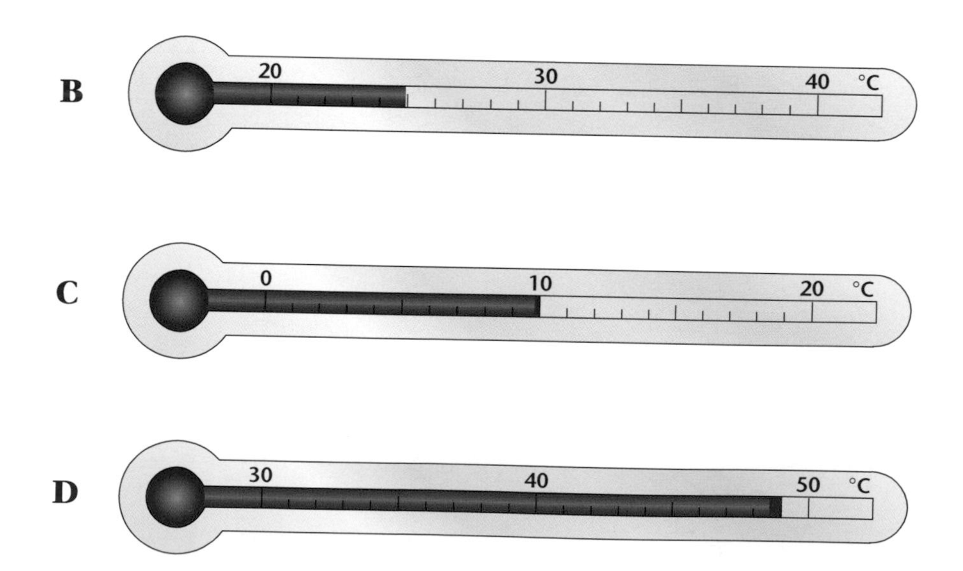

E

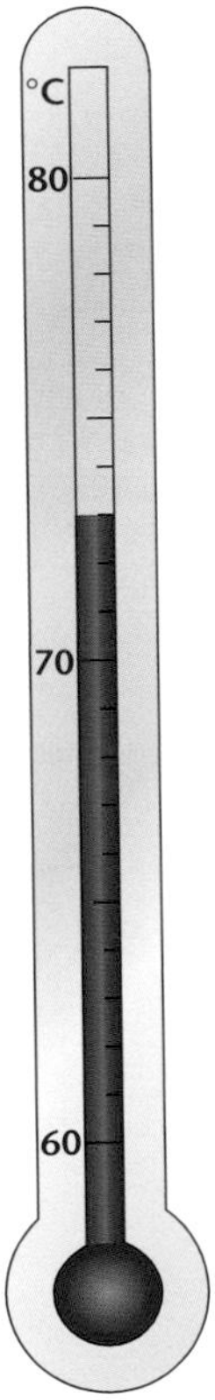

°C
80
70
60

F

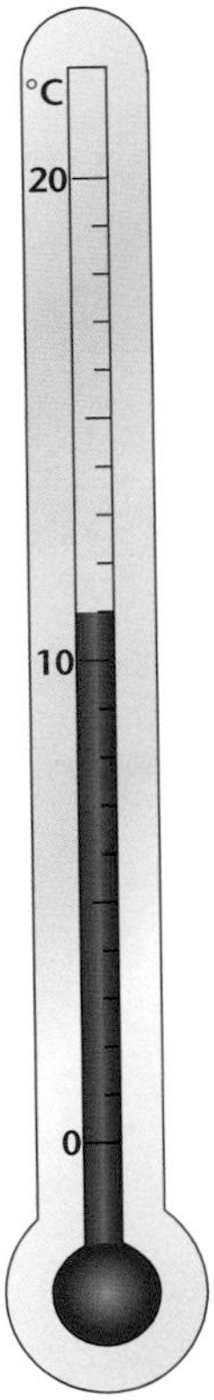

°C
20
10
0

G

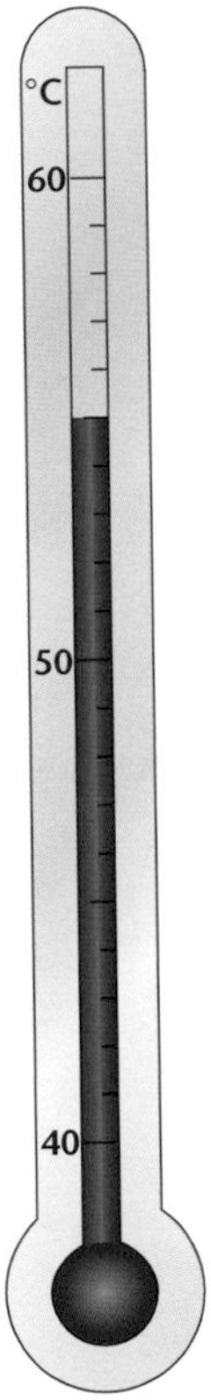

°C
60
50
40

H

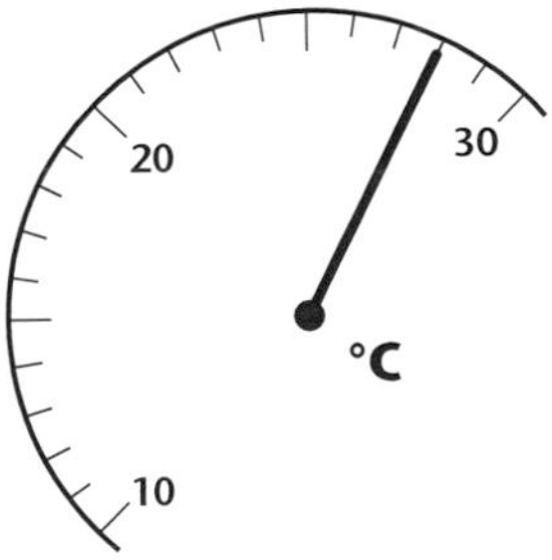

20
30
10
°C

I

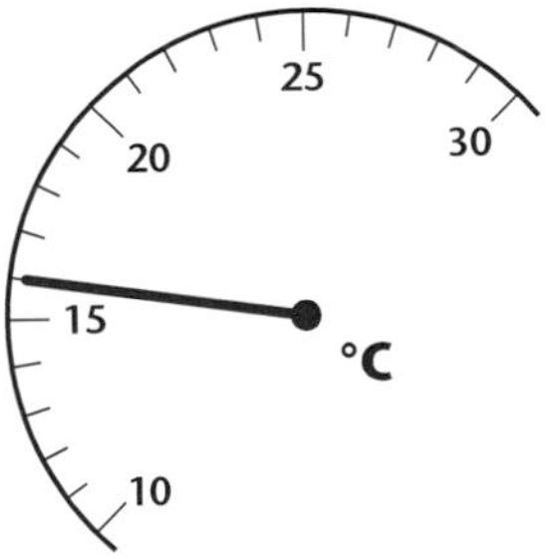

25
20
30
15
10
°C

J

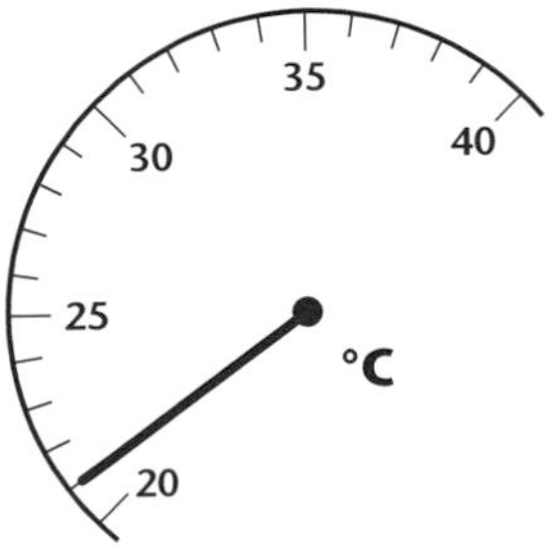

35
30
40
25
20
°C

K

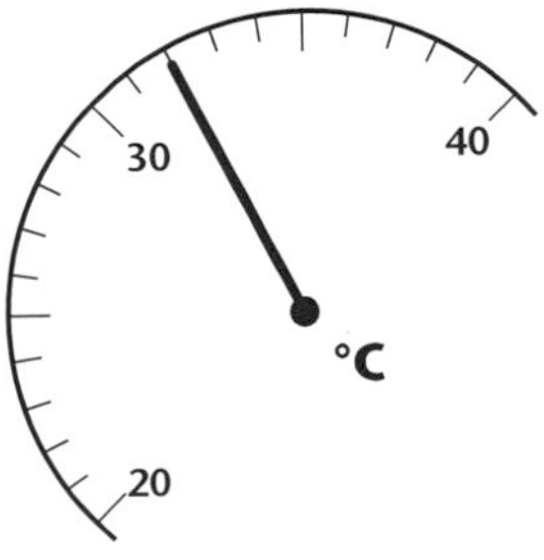

30
40
20
°C

# Clocks 2

The long hand on this clock is pointing to the **6**.

The time is **half past** ☐.

The short hand on this clock is **past** the **7**.

The time is **half past 7**.

► What time does each clock show?

**A**

**B**

**C**

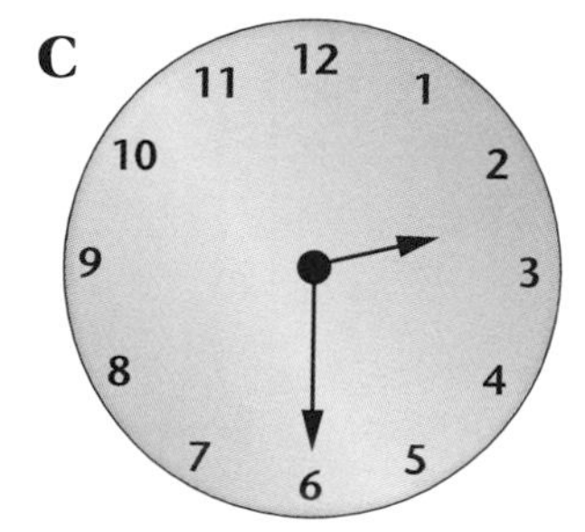

The long hand is pointing to the **3**.

The time is **quarter past** ☐.

The short hand is **past** the **11**.

The time is **quarter past 11**.

► What time does each clock show?

**D**

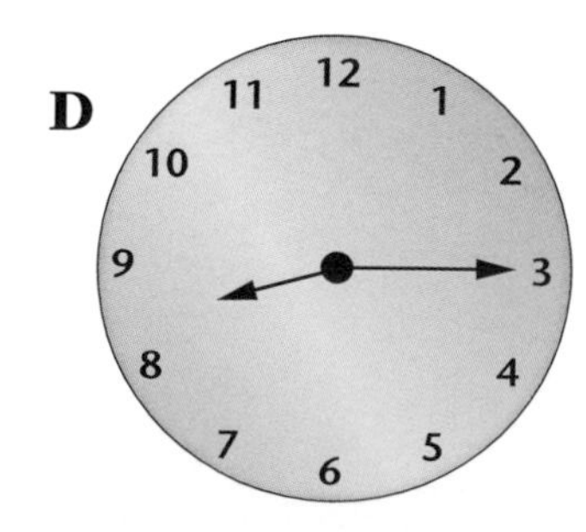

**E**

**F**

The long hand is pointing to the **9**.

The time is **quarter to** ______.

The short hand is **before** the **4**.

The time is **quarter to 4**.

▶ What time does each clock show?

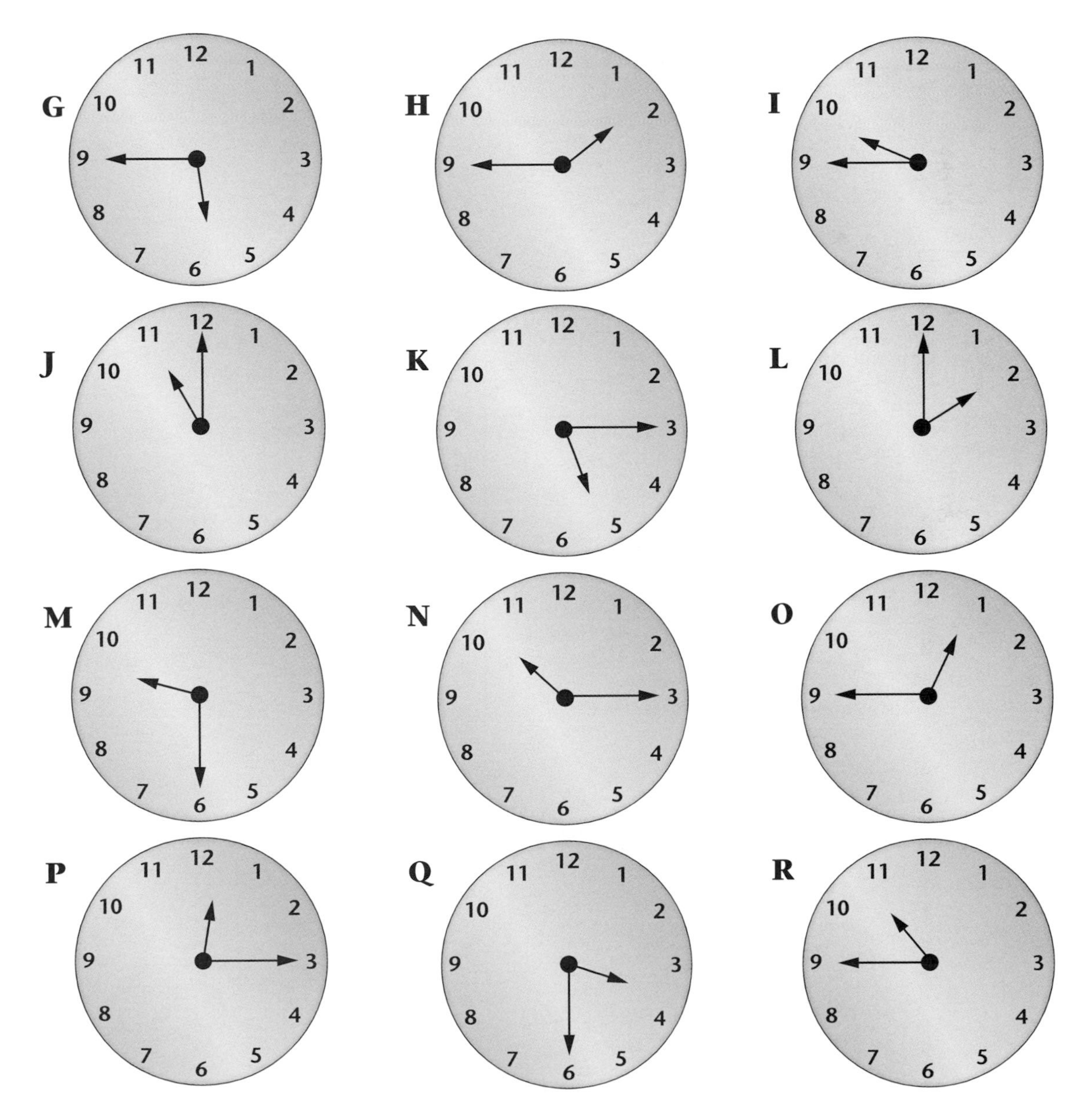

# Ordering money

► Which of these is **more**?

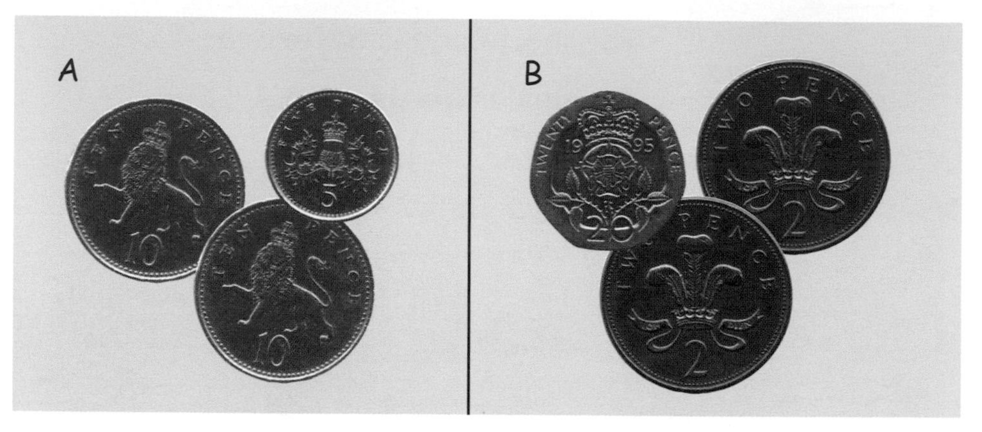


A is 25p

B is 24p

A is more.

► How much money is in each box?
For each pair, write down which is **more**.

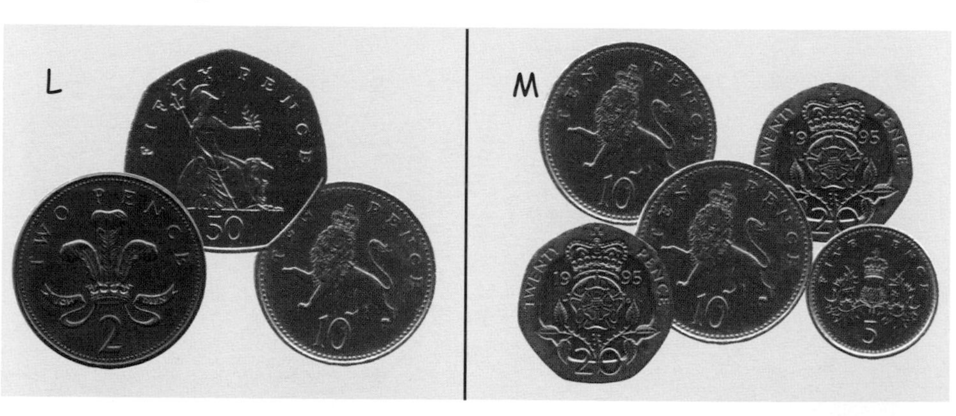

► Work out how much money is in each box.
Write the amounts in order. Start with the **least**.

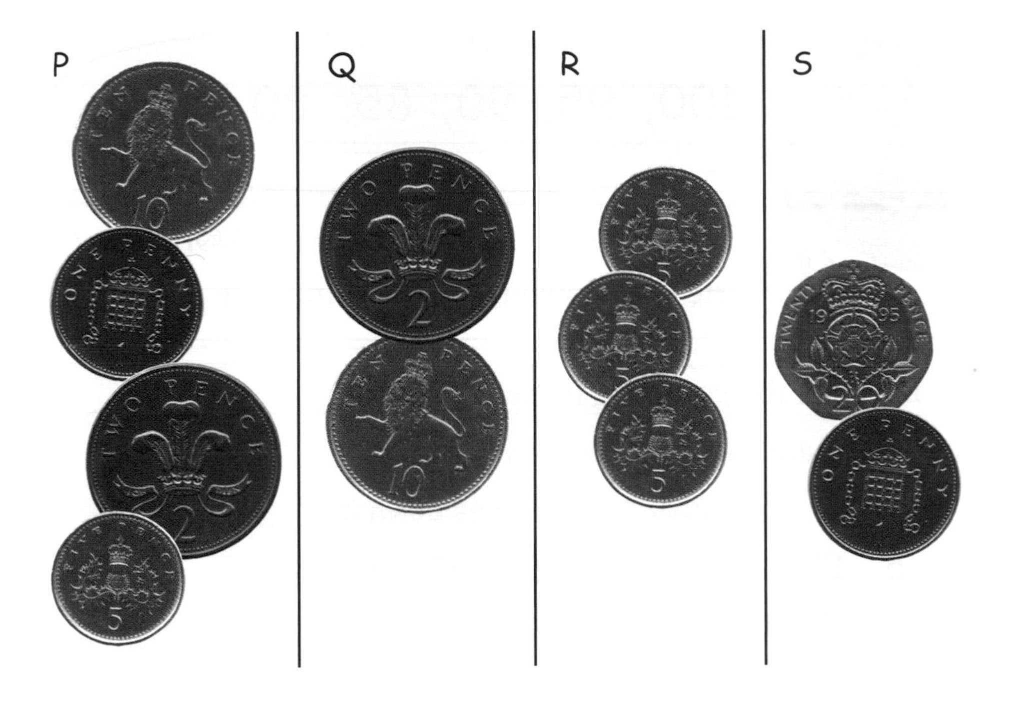


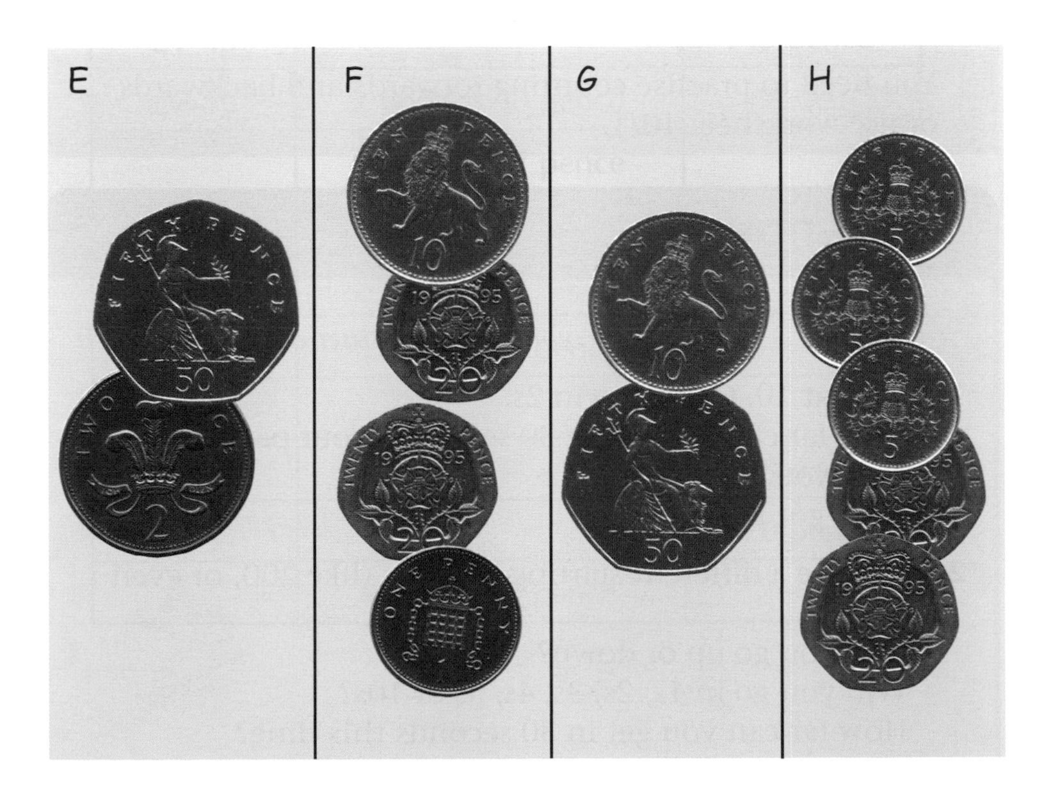


Unit 10 Ordering money

# Unit 12

## Decimals

### Tenths

This rectangle is split into **10** equal pieces.
Each piece is called **one tenth**.

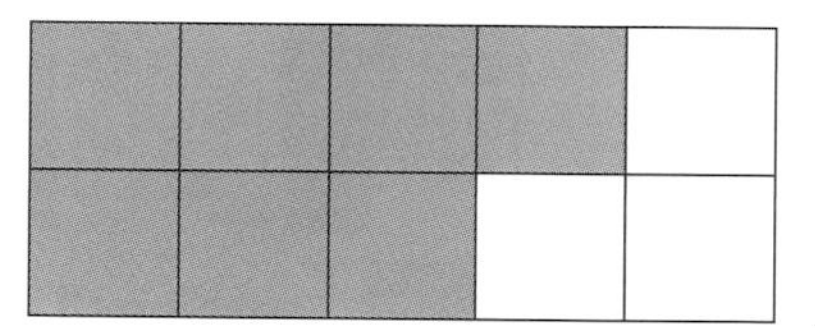

There are 7 pieces shaded out of ten.
This is written **seven tenths**.

▶ Write down the amount shaded in each shape.

**A**

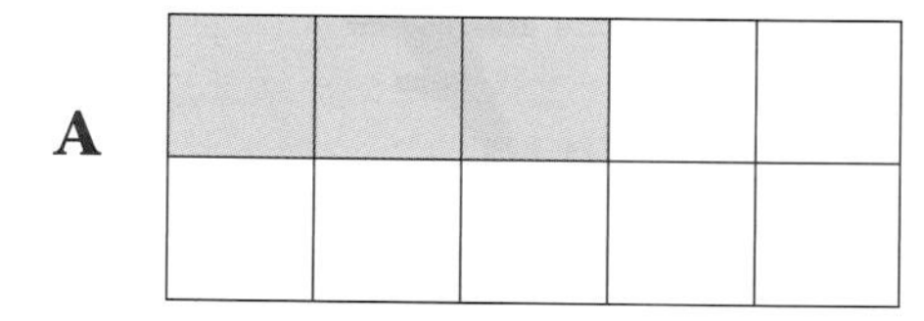

**B**

**C**

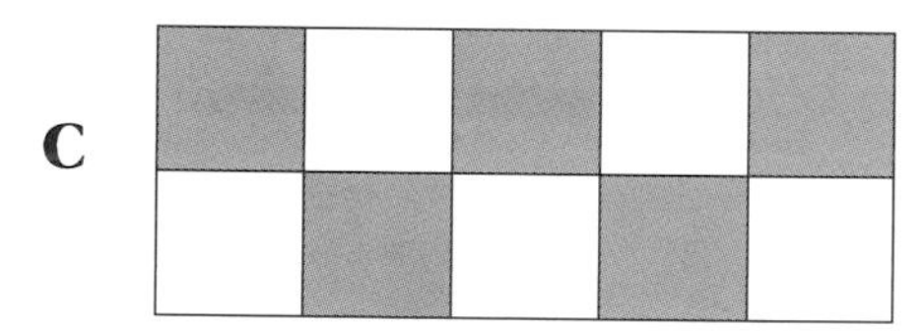

**D**

### Place value

4·8 means 4 units and 8 tenths

You need to practise understanding and writing decimals . . . or use worksheet 12/1.

▶ Copy and complete these sentences.

**A** **7·3** has ___ units and ___ tenths.
**B** **2·6** has ___ units and ___ tenths.
**C** ___ has **4** units and **1** tenth.
**D** **0·9** has ___ units and ___ tenths.
**E** ___ has **8** units and **5** tenths.

## Ordering decimals

These numbers are jumbled up.

0·4 0·7 0·1 0·6

This list has decimal numbers in order.

[0·1] 0·2 0·3 [0·4] 0·5 [0·6] [0·7] 0·8 0·9

The numbers are **in order** in the list.

**smallest** **largest**

0·1 0·4 0·6 0·7

▶ Write each set of numbers in order, from **smallest** to **largest**.
*(. . . use worksheet 12/1 to help you)*

| | | | | |
|---|---|---|---|---|
| **A** | 0·7 | 0·2 | 0·9 | 0·1 |
| **B** | 2·8 | 2·3 | 2·1 | 2·4 |
| **C** | 1·1 | 1·9 | 1·4 | 1·8 |
| **D** | 4·8 | 4·3 | 4·0 | 4·7 |
| **E** | 9·6 | 9·1 | 9·9 | 9·0 |

▶ Write each set of numbers in order, from **largest** to **smallest**.

| | | | | | |
|---|---|---|---|---|---|
| **F** | 3·5 | 3·2 | 3·8 | 3·4 | 3·6 |
| **G** | 7·4 | 7·2 | 7·9 | 7·8 | 7·3 |
| **H** | 5·0 | 5·7 | 5·4 | 5·3 | 5·6 |
| **I** | 8·6 | 8·8 | 8·7 | 8·5 | 8·9 |
| **J** | 6·1 | 6·4 | 6·5 | 6·0 | 6·3 |

# Pictograms 1

Simon asked some people to choose their favourite flavour of ice cream.

He displayed the results in this pictogram.

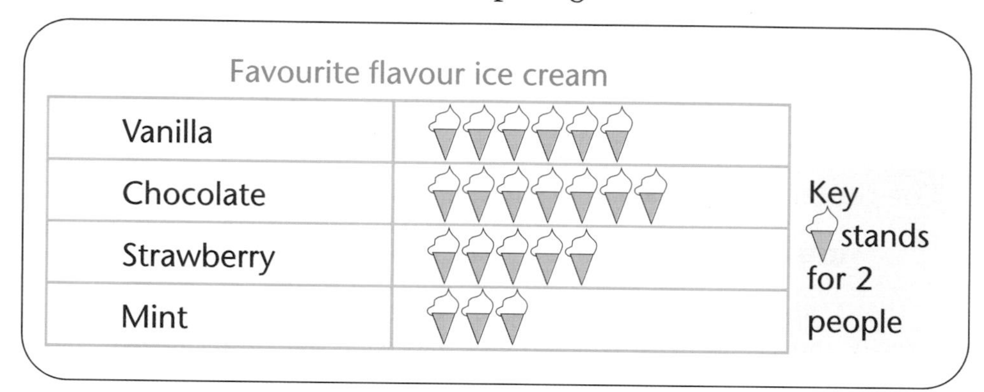


The key to the pictogram shows that [ice cream] stands for 2 people. You can use the pictogram to find out how many people chose each flavour.

Look at the row for vanilla.

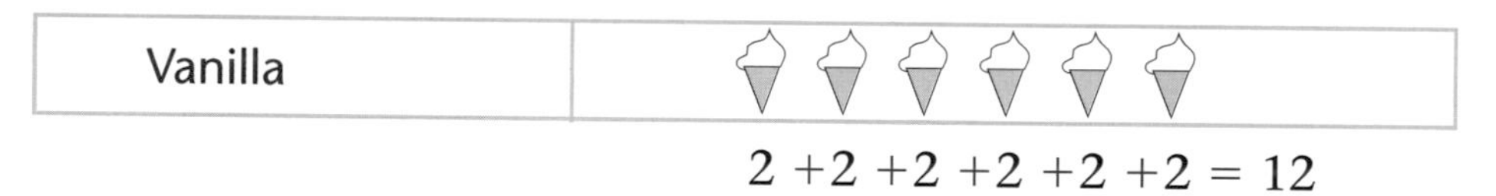


$2 + 2 + 2 + 2 + 2 + 2 = 12$

This shows that 12 people chose vanilla ice cream as their favourite.

► Answer these questions about Simon's pictogram.

**1** How many people chose chocolate?

**2** How many people chose strawberry?

**3** How many people chose mint?

**4** What was the most popular flavour?

**5** How many people did Simon ask altogether?

▶ Amy asked people how long they watched TV for in an evening.
She drew this pictogram of her results.

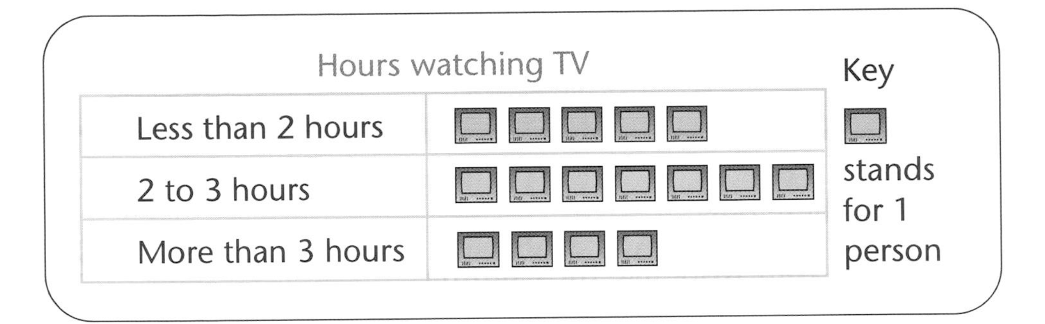


6 How many people watched TV for less than 2 hours?

7 How many people watched TV for more than 3 hours?

8 How many people did Amy ask altogether?

▶ A travel agent used a pictogram to display the holidays sold in one week.

9 What was the most popular holiday?

10 What was the least popular holiday?

11 How many holidays to Florida were sold?

12 How many holidays to Spain were sold?

# Clocks 4

5 to
10 to
quarter to
20 to
25 to

The long hand is pointing to the **10**.

The time is **ten to** ☐.

The short hand on this clock is **before** the **8**.

The time is **ten to 8**.

A digital clock would show **7:50**.

7 o'clock and
50 extra minutes

► What time does each clock show?
Write your answers both ways.

**A**

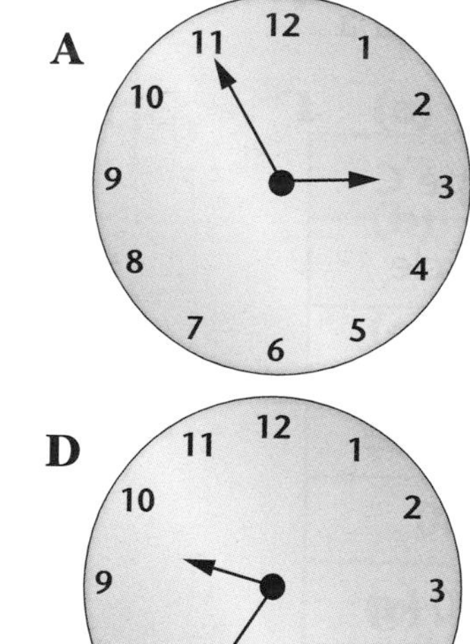

**D**

**B**

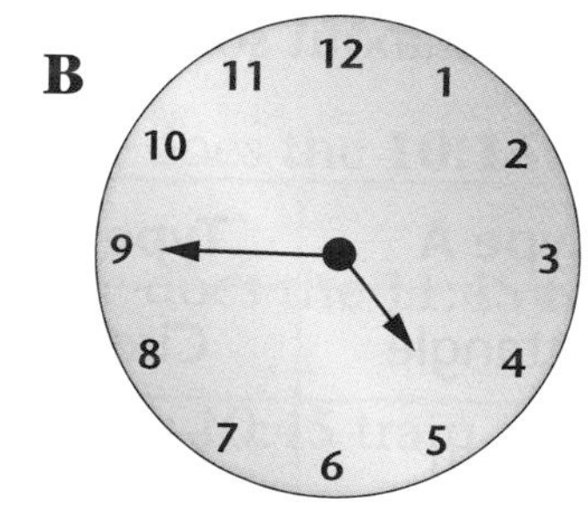

**C**

**E**

**F**

**G**

**H**

**I**

# Frequency tables

▶ Kelly asked the people in her class some questions about sport.

Here are the results of her survey.

| Q1 Which sport do you like best? | |
|---|---|
| **Sport** | **Number of people** |
| Football | 11 |
| Tennis | 8 |
| Athletics | 3 |
| Cricket | 6 |

**A** Which sport did most people like?
**B** Which sport did least people like?
**C** How many people liked athletics?
**D** How many people liked football?
**E** How many people did she ask altogether?

| Q2 Which football team do you like best? | |
|---|---|
| **Team** | **Number of people** |
| Arsenal | 7 |
| Chelsea | 2 |
| Aston Villa | 4 |
| Man United | 10 |
| Newcastle | 5 |

**F** Which football team did most people like?
**G** Which football team did least people like?
**H** How many people liked Newcastle?
**I** Which team did 7 people like?
**J** How many people liked Aston Villa?

You need to practise making frequency tables . . . or use worksheet 13/3.

# Mirrors 3

This is **part** of a reflection pattern. This is the **completed** pattern.

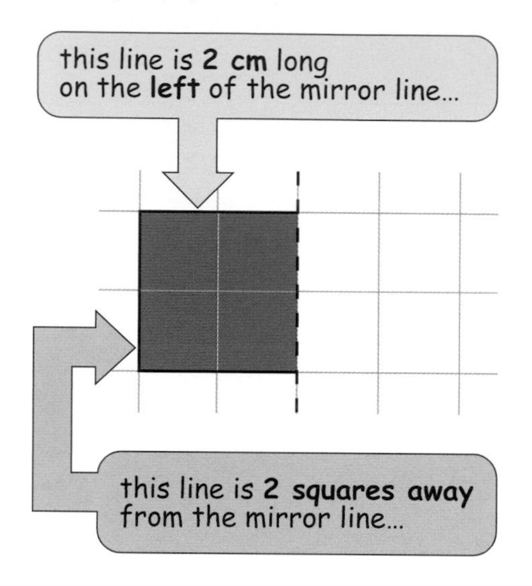


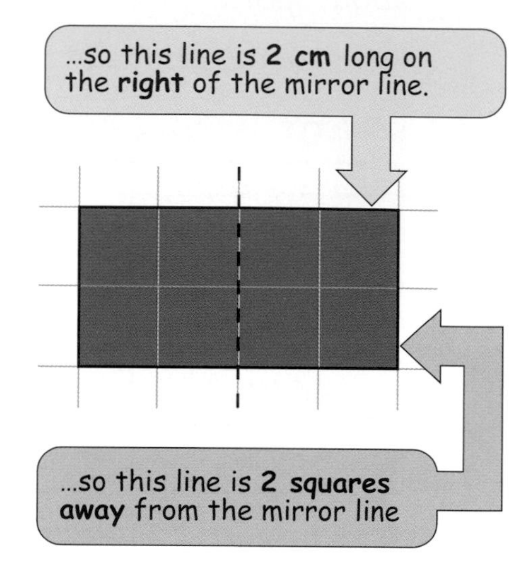


► Which of these reflection patterns have been completed correctly?

Write **yes** or **no**.

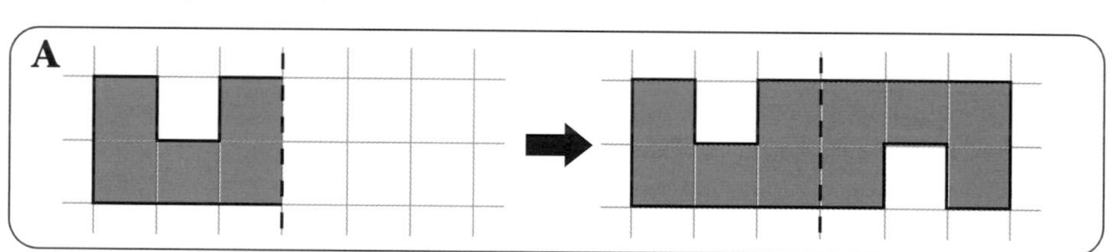


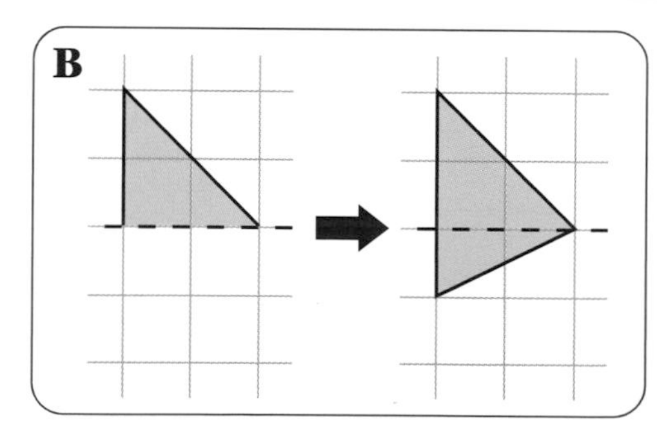


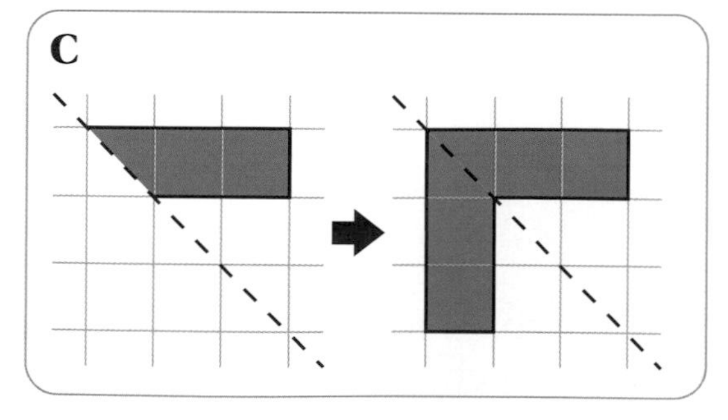


You need to practise completing reflection patterns . . . or use worksheet 13/4.

# Lines 4

## Measuring lines

This line is 32 millimetres long.

This line is 9 millimetres long.

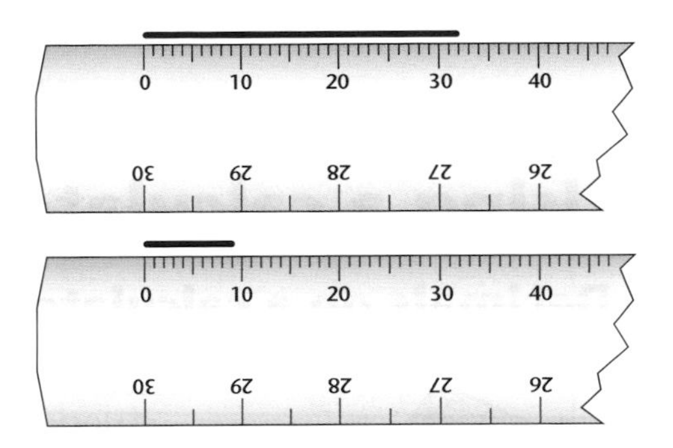

 You need to practise measuring lines in millimetres . . . or use worksheet 13/5.

## Drawing lines

► To draw a line 25 mm long:

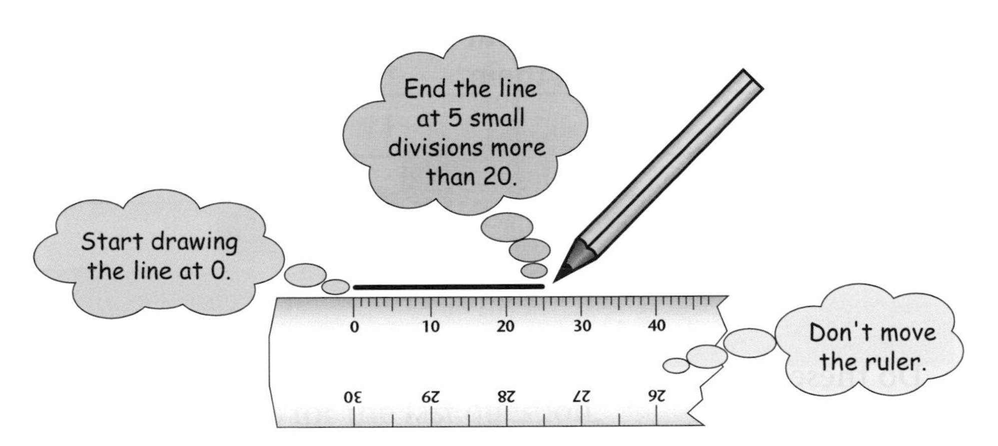


► Draw these lines.

Make them as accurate as you can.

| | | | | | |
|---|---|---|---|---|---|
| **A** | 52 mm | **E** | 49 mm | **I** | 63 mm |
| **B** | 38 mm | **F** | 20 mm | **J** | 84 mm |
| **C** | 15 mm | **G** | 76 mm | **K** | 100 mm |
| **D** | 7 mm | **H** | 25 mm | **L** | 9 mm |

# Lines and angles

## Review: right angles

▶ Which of these angles are right angles?
Make a right angle from cubes to help you.

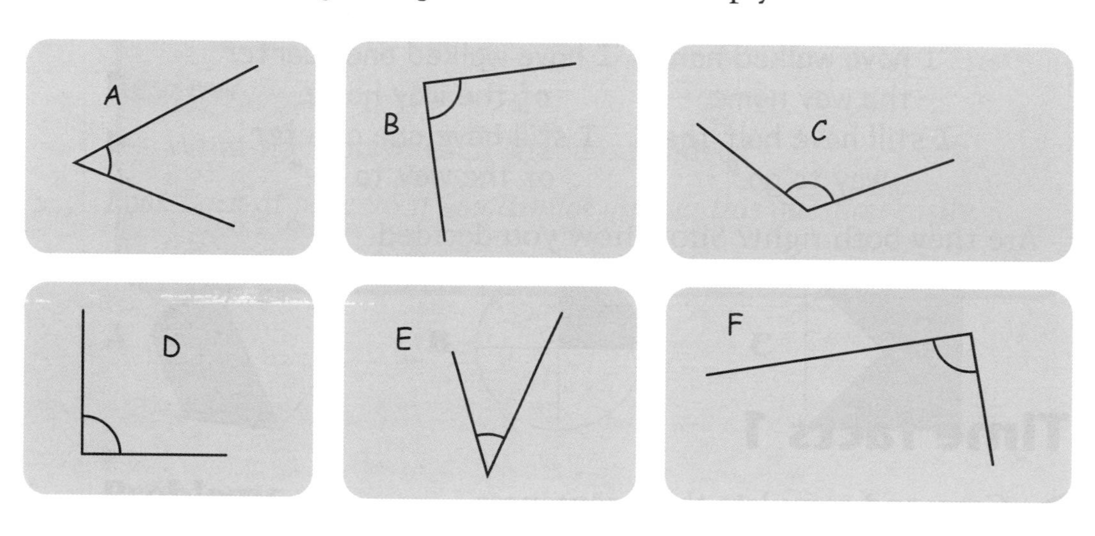


## Parallel lines

Parallel lines

Parallel lines never meet.
The lines are always the same distance apart.

1 Which pairs of lines are parallel?

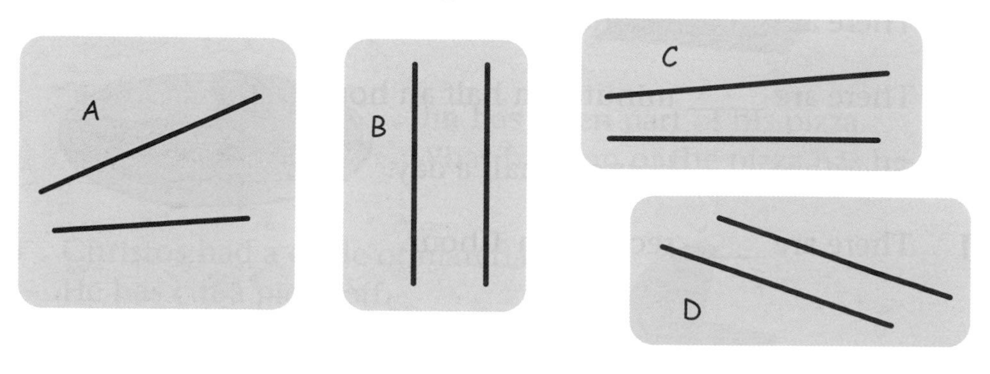

**2** Which lines are parallel to line **E**?

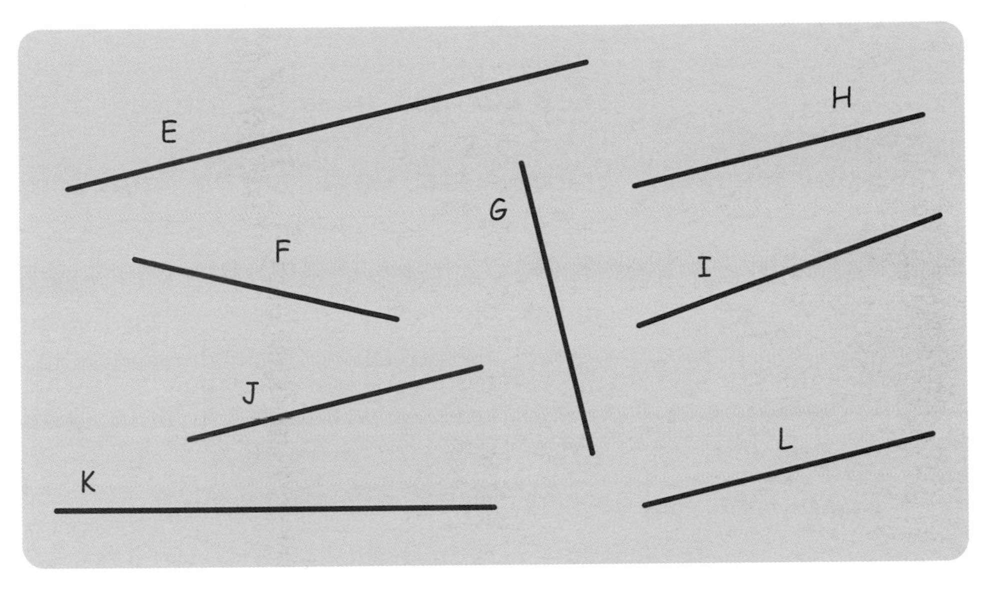


**3** Which lines are at right angles to line **M**?

**4** Which lines are parallel to line **M**?

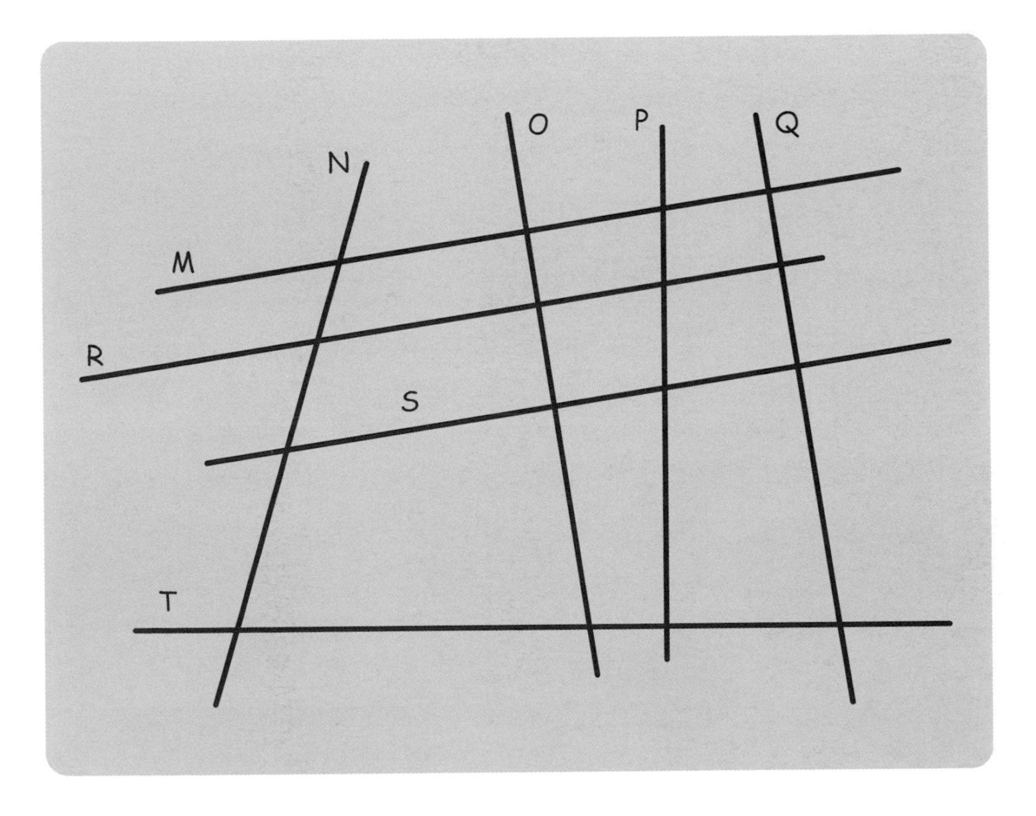

# Shapes 4

This is a **square**.
The **4 corners** are **right angles**.
The **4 edges** are the **same length**.

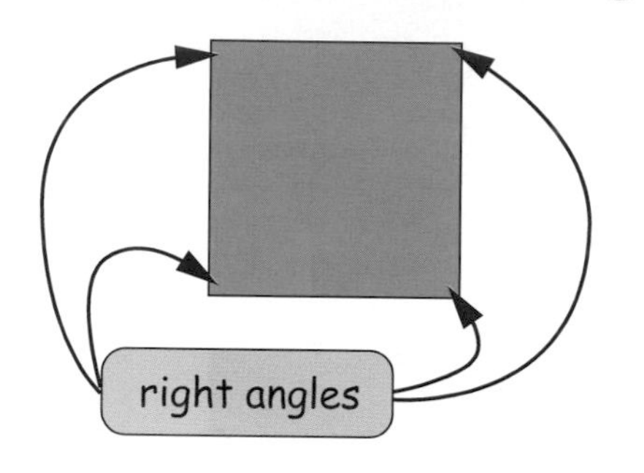


This is a **rectangle**.
The **4 corners** are **right angles**.
The **4 edges** are not all the same.
The **opposite edges** are the **same length**.

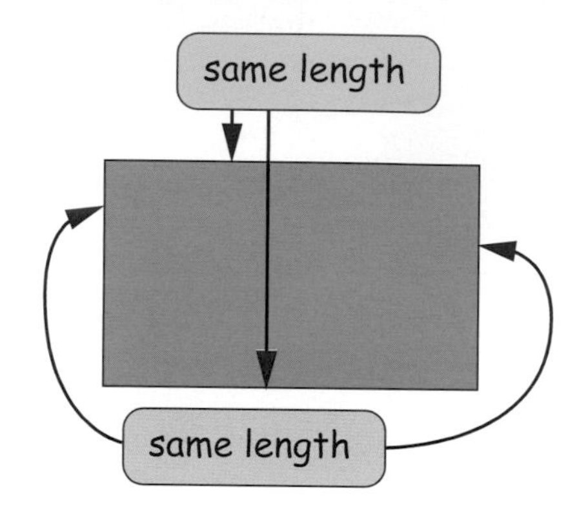


- Copy this table.
- Look at the shapes below. Put the letters in the correct place in the table.

| square | rectangle | neither |
|---|---|---|
| | | |

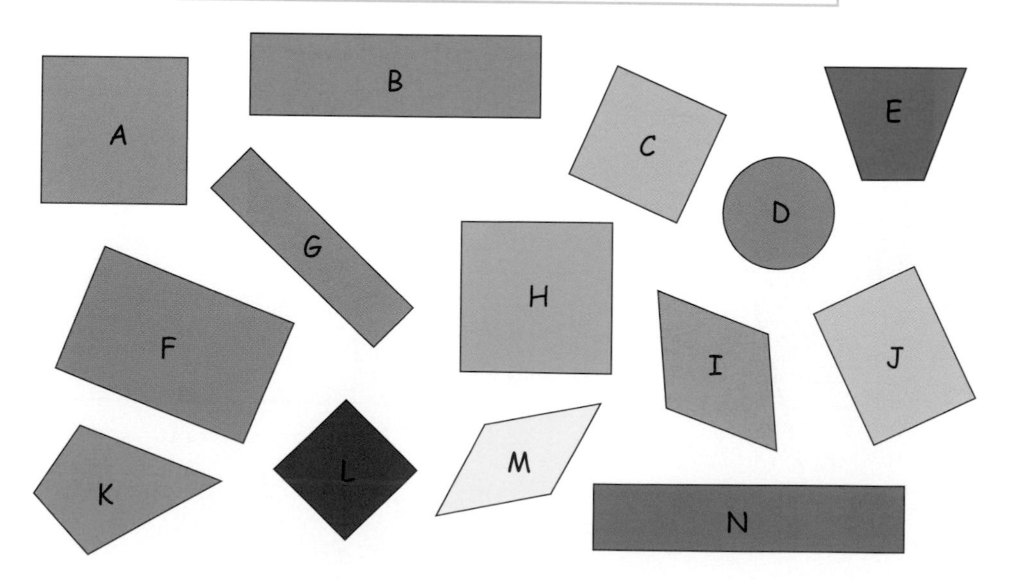

# Money 4

▶ Make these amounts of money from the coins above.
Write down which coins you use each time.

**A** £2·60

**B** £1·45

**C** £3·80

**D** £2·05

**E** £4·23

**F** 89p

**G** £3·72

**H** £1·99

**I** £2·36

**J** £4·07

**K** 67p

**L** £3·14

▶ Count up the money in the picture.
How much money is there altogether?

You need to practise counting out amounts of money . . .
or use the Money Cards.

# Add and subtract 5

line up the digits

adding:
carry the tens digit

always start from the right-hand side

subtracting:
remember to borrow

## Section A

 You can use worksheet 15/1 to start these sums.

**A** 253 + 614

**B** 385 + 760

**C** 385 + 462

**D** 364 + 218

**E** 593 + 264

**F** 485 + 176

**G** 395 + 408

**H** 196 + 357

**I** 294 + 21

**J** 497 + 253

## Section B

 You can use worksheet 15/1 to start these sums.

**K** 584 − 163

**L** 951 − 325

**M** 468 − 193

**N** 273 − 148

**O** 527 − 293

**P** 885 − 291

**Q** 627 − 155

**R** 480 − 327

**S** 508 − 143

**T** 397 − 354

**U** 964 − 380

**V** 574 − 61

# Quarters 1

*you may use cubes or counters*

Quarter always means share into 4 equal groups.

A quarter of 20 means share 20 into 4 equal groups.

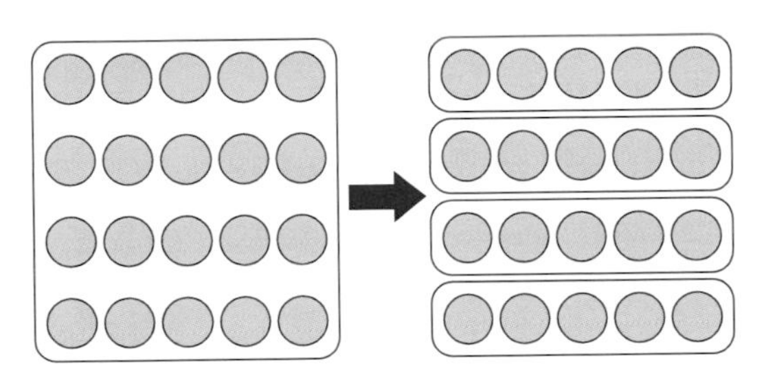

There are 5 in each group.

So a quarter of 20 = 5

▶ Answer these questions.

*Look back at page 106 for the 4 times table.*

**A** a quarter of 12

**B** $\frac{1}{4}$ of 40

**C** a quarter of 8

**D** $\frac{1}{4}$ of 4

**E** a quarter of 28

**F** $\frac{1}{4}$ of 16

**G** a quarter of 20

**H** $\frac{1}{4}$ of 32

**I** Sarah bakes 24 biscuits.
She eats a quarter of them.
How many does she eat?

**J** Erika earns £36.
She saves a quarter of it.
How much does she save?

**K** Steve has 12 new CDs.
He has listened to a quarter of them.
How many has he listened to?

**L** Daljit works for 36 hours a week.
He spends a quarter of the time filing.
How many hours is this?

# Clocks 5

o' clock
5 past
10 past
quarter past
20 past
25 past
half past
25 to
20 to
quarter to
10 to
5 to

The long hand is on the **5** and the short hand is after the **9**.

The time is **25 past 9**.

A digital clock would show **9:25**.

► What time does each clock show? Write your answers both ways.

**A**

**B**

**C**

**D**

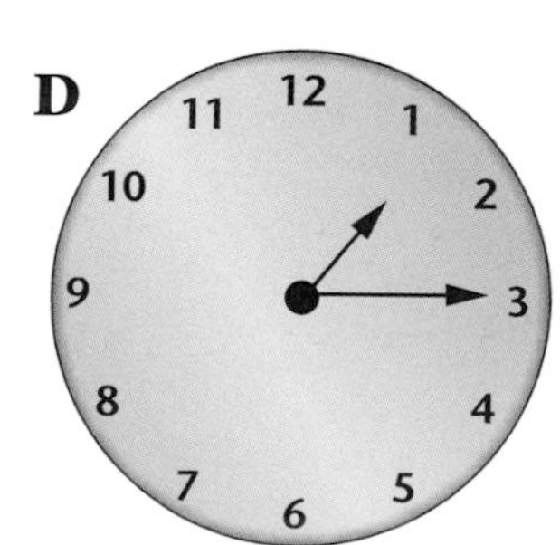

**E**

**F**

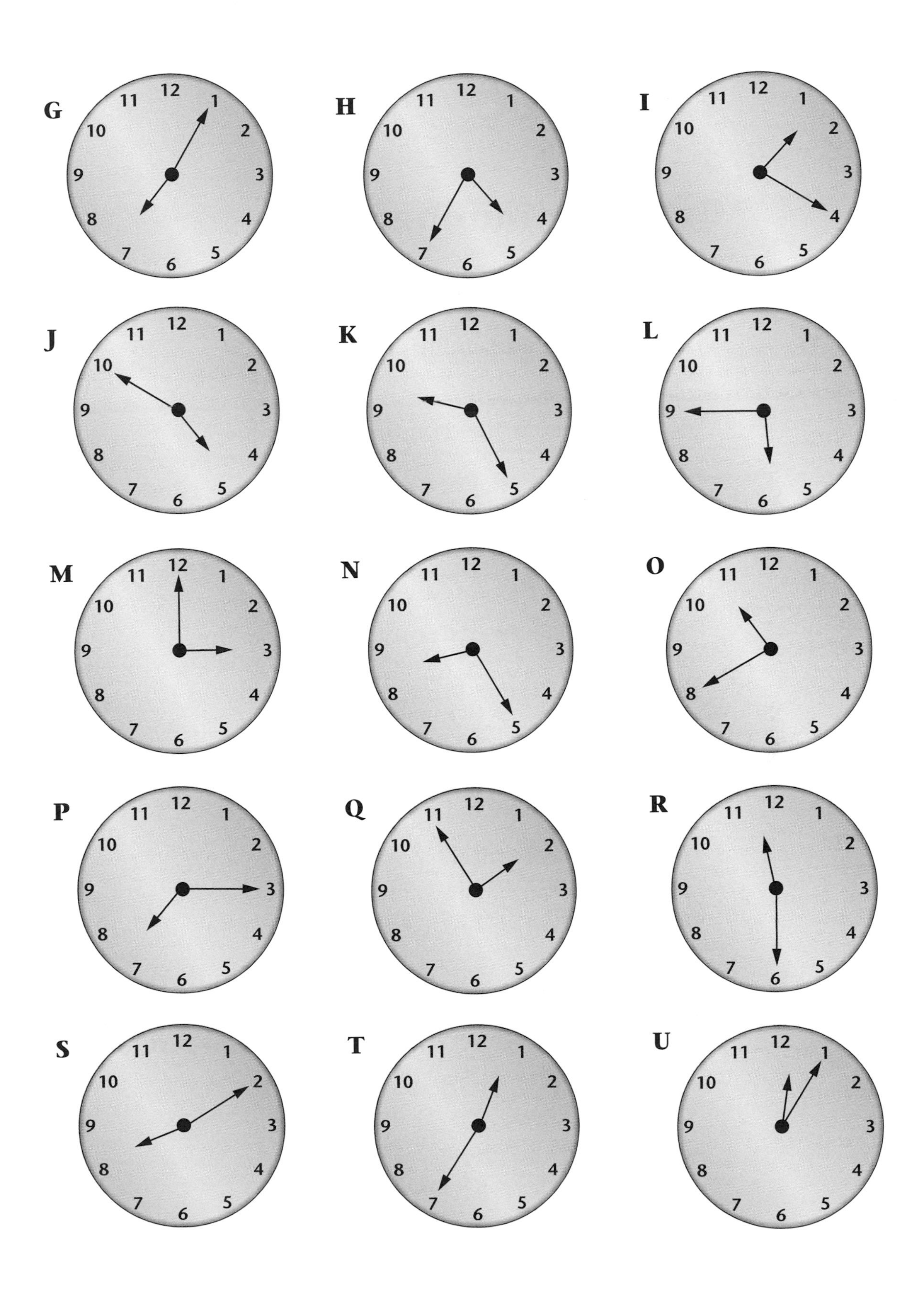
G
H
I
J
K
L
M
N
O
P
Q
R
S
T
U

# Unit 16

## Ordering numbers 2

### Using number lines

These numbers are jumbled up. 203 175 157 230 196

They have been marked on this number line.

157 175 196 203 230

130 140 150 160 170 180 190 200 210 220 230 240 250

The numbers are **in order** on the number line.

**smallest** ... **largest**

157 175 196 203 230

▶ Write each set of numbers in order, from **smallest** to **largest**.

Use a number line to help you.

| | | | | | |
|---|---|---|---|---|---|
| **A** | 360 | 304 | 352 | 325 | 340 |
| **B** | 951 | 927 | 984 | 965 | 948 |
| **C** | 108 | 158 | 110 | 137 | 124 |
| **D** | 282 | 301 | 297 | 352 | 318 |
| **E** | 483 | 834 | 348 | 843 | 384 |

▶ Write each set of numbers in order, from **largest** to **smallest**.

| | | | | | |
|---|---|---|---|---|---|
| **F** | 680 | 698 | 627 | 675 | 643 |
| **G** | 320 | 302 | 326 | 362 | 360 |
| **H** | 284 | 342 | 560 | 729 | 602 |
| **I** | 250 | 25 | 520 | 52 | 205 |
| **J** | 452 | 425 | 254 | 524 | 542 |

## In order

**1** Year 11 are having a raffle for charity.

Here are the numbers of tickets sold by each class.

| 11C | 11F | 11M | 11P | 11R |
|---|---|---|---|---|
| 265 | 189 | 258 | 361 | 309 |

**(a)** Which class sold the most tickets?

**(b)** Which class sold the least tickets?

**(c)** Write the numbers in order. Start with the most.

**2** These are the distances of some places from London.

| York | Glasgow | Holyhead | Dover | Penzance |
|---|---|---|---|---|
| 340 km | 658 km | 452 km | 126 km | 499 km |

**(a)** Which place is nearest to London?

**(b)** Which place is furthest from London?

**(c)** Write the distances in order. Start with the nearest.

**3** These are the heights of the people in the Jones family.

Write their heights in order. Start with the tallest.

# Weighing

These are all from weighing scales.
Check that you understand each scale.

▶ What weight does each scale show?

**A**

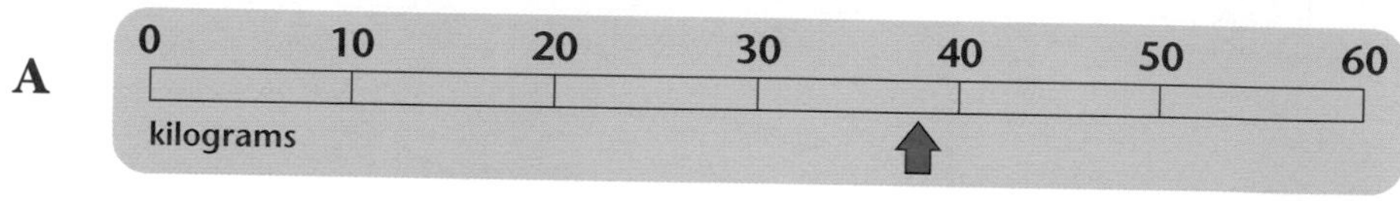


Choose the best answer:

35 kg     36 kg     37 kg     38 kg     39 kg

**B**

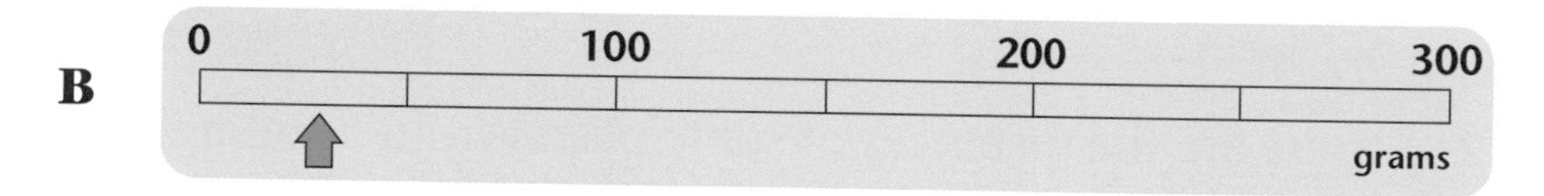


**C**

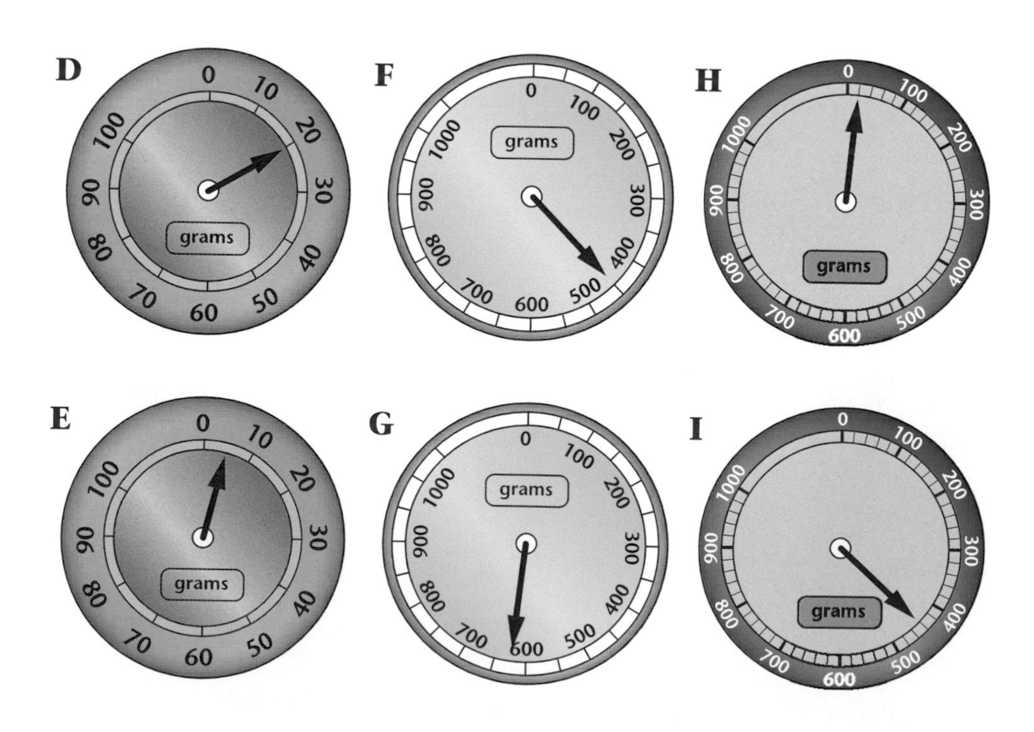

**J** **K** **L**

## Weighing scales

*you need the weighing scales, the objects and a partner*

**1** Before you start:

- ▶ are the scales flat on the table?
- ▶ is the bowl fixed firmly on the top?
- ▶ is the pointer at zero?
- ▶ is the item you are weighing too big for the bowl?

**2** Choose at least 5 items.
Weigh each one as accurately as you can.

Write the weight in your book like this:

box of counters = __________ grams

# Pentagons

*you need square dotty paper*

A pentagon is a shape with five sides.

How many **different** pentagons can you draw on one sheet of paper?

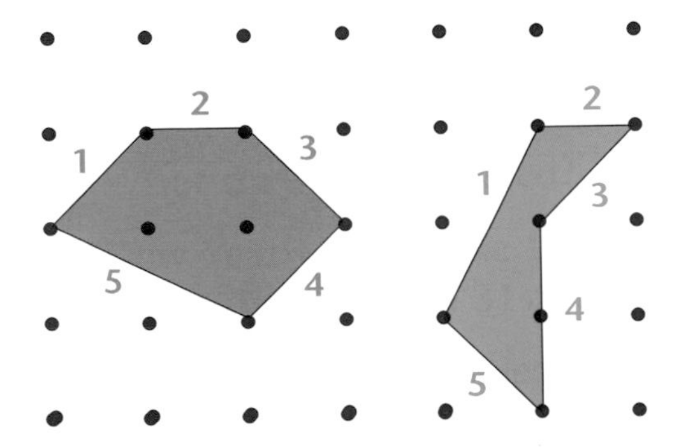


You could try other shapes. . .
hexagons . . .
octagons . . .

# Dividing 2

multiply times divide share
double twice half

▶ Each shape in these sums stands for a different number.
Copy and complete these sums.

**A** 2 × ◆ = 20    20 ÷ 2 = ◆    20 ÷ ◆ = 2

**B** ● × 7 = 21    21 ÷ ● = 7    21 ÷ 7 = ●

**C** 6 × 5 = ✱    5 × 6 = ✱    ✱ ÷ 5 = 6

**D** 7 × ✸ = 14    14 ÷ ✸ = 7    ▲ ÷ 7 = 2

**E** ■ × 4 = 24    24 ÷ ■ = 4    24 ÷ ★ = 6

**F** ✦ × 1 = 4    4 ÷ ✿ = 4    4 ÷ ✦ = 1

▶ Copy and complete these sums.

**G** three times six equals ____

____ shared by three equals six

**H** four multiplied by ____ equals forty

forty divided by four equals ____

**I** twice seven equals ____

half of ____ equals seven

**J** eight times five equals ____

____ shared by five equals eight

**K** double nine equals ____

half of ____ equals nine

**L** four multiplied by ____ equals sixteen

sixteen divided by four equals ____

▶ Answer these questions.

**M** Dan has ten marbles.
He shares them equally with his brother.
How many do they get each?

**N** Bill, Ben and Wendy win £12.
They share the money equally.
How much do they get each?

**O** 

Toby has a piece of material 8 m long.
He needs 2 m to make a curtain.
How many can he make?

**P** Sivay buys five candles.
They cost £15 altogether.
How much do they cost each?

**Q** Ainsley makes 25 biscuits.
He shares them equally with four friends.
How many do they get each?

**R** Ted needs 20 hooks.
They are sold in packs of 5.
How many packs must he buy?

**S** Rita, Sue and Bob buy a bunch of flowers for Mother's Day.
It costs £9 altogether.
They share the cost equally.
How much should they each pay?

**T** Anita has 6 samosas.
She shares them equally with her brother.
How many do they get each?

# Change 4

## Change from £5

One way to make £5 is like this.

How much change do you get from £5 if you spend £2·70?

spend

£2·70

change

£2·30

► Work out how much change you get from £5 if you spend these amounts.
Use coins to help you.

**A** £3·60

**B** £2·40

**C** £1·20

**D** £4·80

**E** £0·70

**F** £1·90

**G** £3·10

**H** £4·50

**I** £3·30

**J** £2·90

**K** £1·60

**L** £0·80

**M** £4·10

**N** £3·20

# Unit 17

## Times and divide

▶ Work out

**A** 16 multiplied by 4

**B** 75 shared by 5

**C** 651 divided by 7

**D** 29 times by 8

**E** 52 shared by 13

**F** 108 multiplied by 3

**G** 63 times by 12

**H** 852 divided by 4

▶ These questions are all about Cara's cafe.

**I** There are 12 tables. Each table has 6 chairs.
How many chairs are there altogether?

**J** Cara makes 15 scones in one batch.
How many batches should she make if she needs 180 scones?

**K** 

Each jar of jam makes 20 portions.
How many jars does she need for 180 portions of jam?

**L** Cara uses 36 pints of milk each day.
How many pints of milk does she use in 6 days?

**M** Each pot of coffee fills 9 cups. She makes 22 pots of coffee one day.
How many cups of coffee has she filled?

**N** Cara cuts a chocolate cake into 12 slices.
How many cakes does she need for 108 slices?

# Unit 18

## Using decimals

**1**

| Price list | |
|---|---|
| Adults | £2·49 |
| Children | £1·49 |
| OAPs | £1·75 |

**(a)** How much is it for 2 adult tickets?

**(b)** How much is it for 1 adult and 1 child?

**(c)** How much more does an adult ticket cost than an OAP ticket?

**2** John is 1·60 m tall. Chris is 1·82 m tall.

**(a)** Who is taller?

**(b)** How much taller is he?

**3** What is the total height of the table and the lamp?

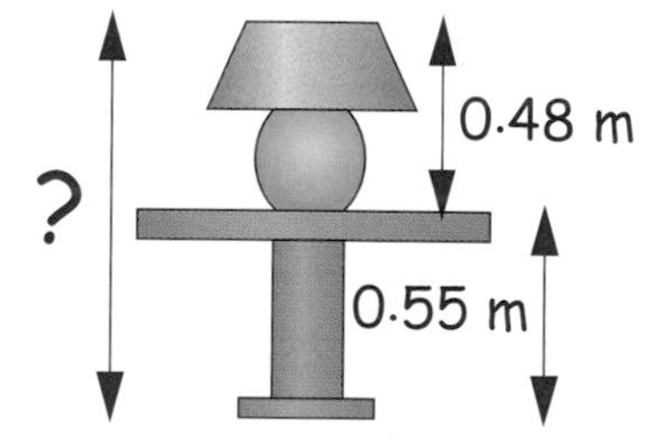


**4** Lou has a 1·5 kg bag of flour.
He uses 0·6 kg.
How much does he have left?

**5**

Coral buys 0·75 kg of apples, 1·2 kg of bananas and 0·45 kg of grapes.
What does her fruit weigh altogether?

**6** What is the height of the wardrobe door?

2·36 m

?

0·80 m

**7** 

Baby Ashia weighed 3·76 kg when she was born.
At 6 months old she weighed 8·24 kg.
How much weight has she gained in 6 months?

**8** Emily buys a burger and fries.
How much does it cost?

**9** Liam buys chicken pieces and a shake.
How much does it cost?

**10** Neil buys a burger, fries and a coke.

**(a)** How much does it cost?

**(b)** He pays with a £5 note.
How much change does he get?

| MENU | |
|---|---|
| Burger | £1·25 |
| Chicken pieces | £1·40 |
| Fries | £0·99 |
| Shake | £1·45 |
| Coke | £1·29 |

**11** 

Parcels sent second class cannot be heavier than 0·75 kg.
Nasim has a parcel that weighs 0·813 kg.
How much heavier than 0·75 kg is Nasim's parcel?

# Reading scales 4

This scale goes up in **ones**.

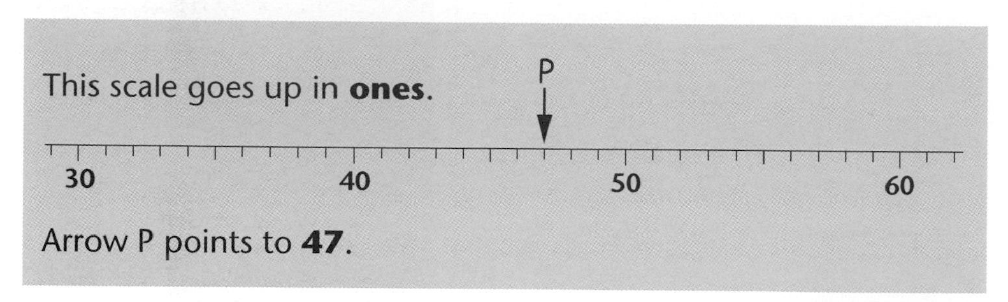


Arrow P points to **47**.

This scale does **not** go up in **ones**.

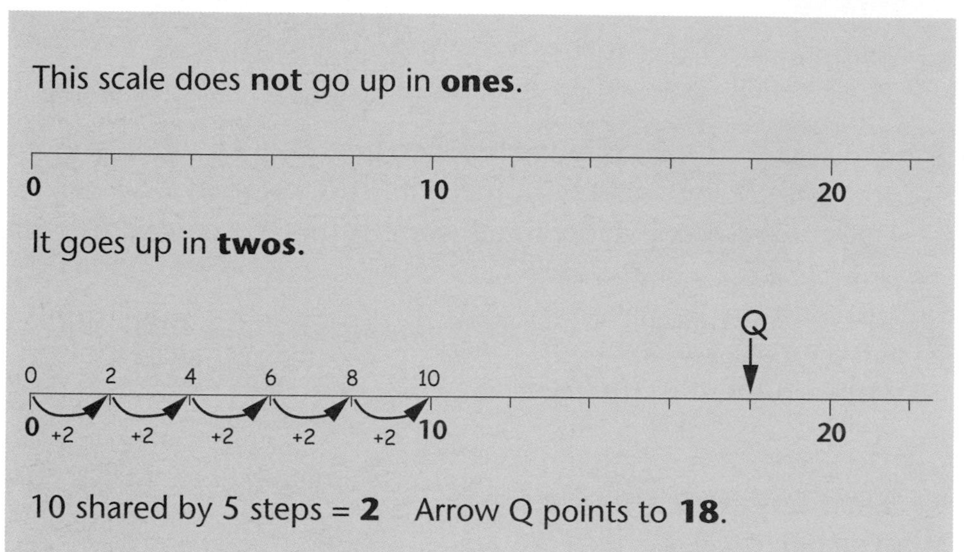


It goes up in **twos**.

10 shared by 5 steps = **2** Arrow Q points to **18**.

This scale goes up in **fives**.

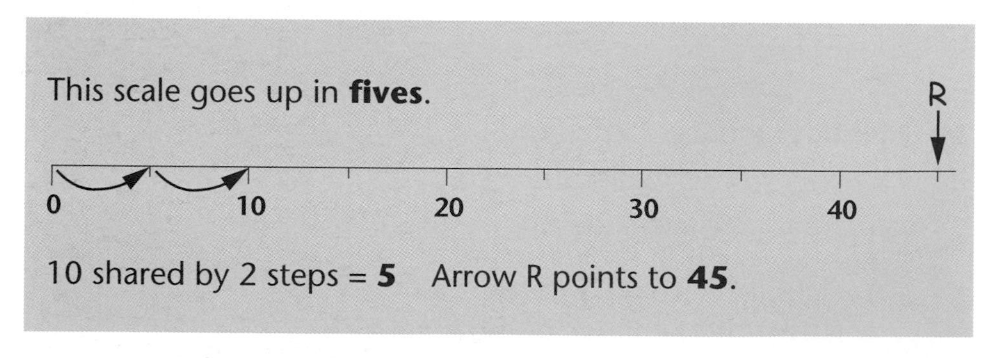


10 shared by 2 steps = **5** Arrow R points to **45**.

You need to practise marking scales that go up in 2s, 5s and 10s . . . or use worksheet 18/1.

▶ Write down the number each arrow points to.

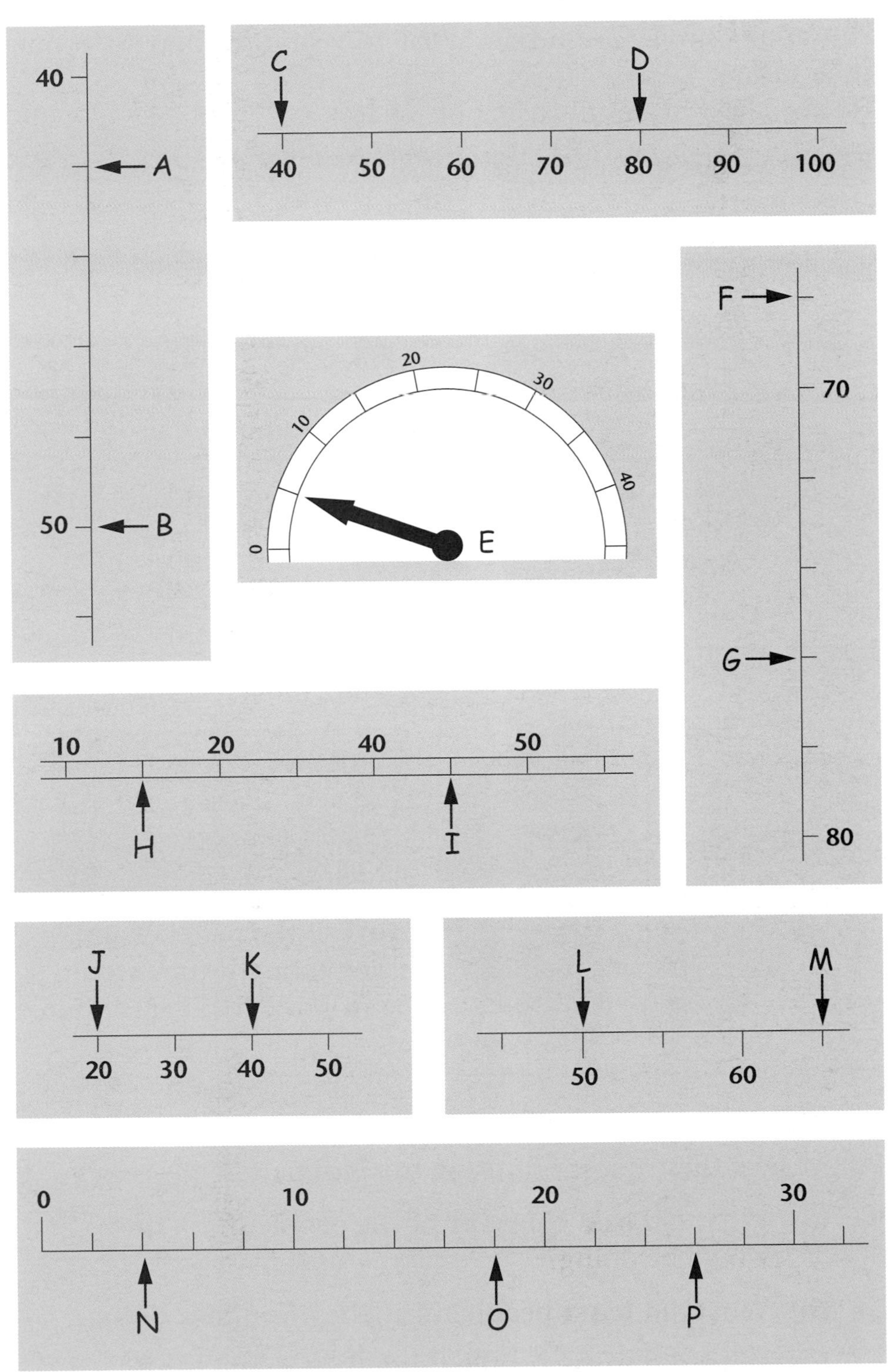


Unit 18 Reading scales 4

# Pie charts

Zahir did a survey to find out what 16-year-olds thought about their future.
He displayed his results using pie charts.

Here are the results of his first question.

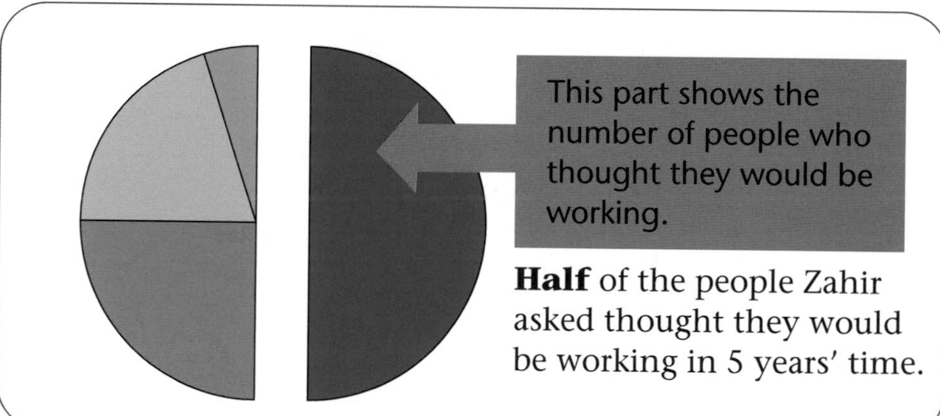


**Half** of the people Zahir asked thought they would be working in 5 years' time.

► Answer these questions using the pie charts.

**1** **(a)** What did **one quarter** of the people think they would be doing?

**(b)** What did **least** people think they would be doing?

**(c)** What did **most** people think they would be doing?

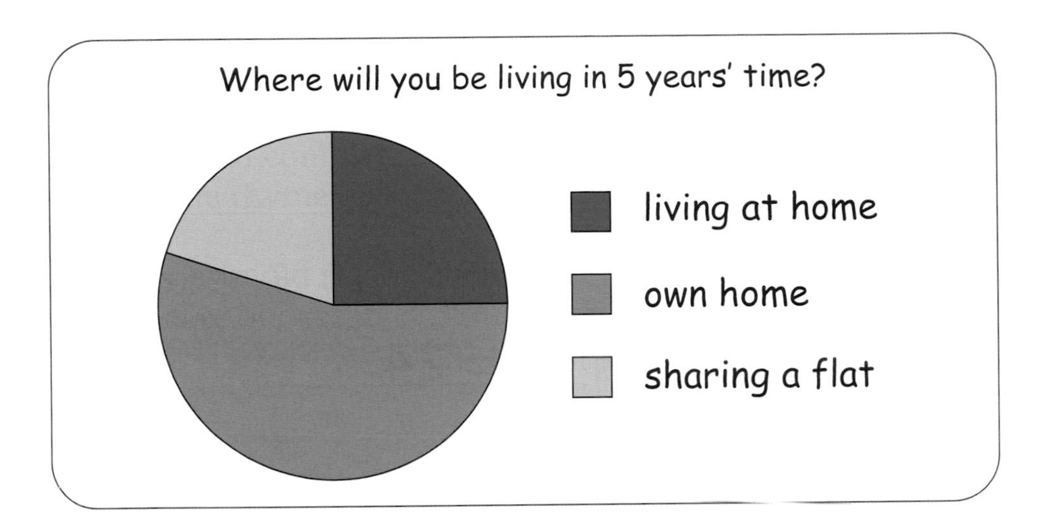


2 **(a)** Where did most people think they would be living?

**(b)** Where did least people think they would be living?

**(c)** What fraction of people thought they would be living at home?

3 **(a)** What fraction of the people said they would be single?

**(b)** What fraction of the people said they would be with a partner?

**(c)** What was the most popular answer?

**(d)** Write the answers in order, from the most popular to the least popular.

# Using l and ml

1 There are 18 litres of petrol in Cameron's tank.
He puts another 24 litres in.
How many litres are in the tank altogether?

2 Suzy buys a 650 ml can of oil. She uses 125 ml.
How much is left in the can?

3 

The bottle holds 325 ml.
Mandy pours 174 ml into some glasses.
How much is left in the bottle?

4 This is a recipe for salty lassi.
Bob wants to make twice as much.
Write down the recipe Bob should use.

Salty Lassi
400 ml plain yogurt
1·4 litres ice-cold water
1 teaspoon salt

5 Lemonade is sold in 3 litre bottles.
I buy 6 bottles for a party.
How many litres is that altogether?

6 

This carton holds 500 ml of milk. I use 125 ml.
How much milk is left in the carton?

7 Tina needs 5 litres of paint to paint her bedroom.

The paint is sold in these tins.
Which size tins should she buy?

# Unit 19

## 24 hour clock 2

05:05 means 5:05 am
or five past five in the morning

17:05 means 5:05 pm
or five past five in the afternoon

► Match the right word to each of these times.

morning | afternoon | evening | night

**A** 15:30

**B** 21:05

**C** 04:10

**D** 00:45

**E** 12:40

**F** 17:50

► Write these as **am** or **pm** times.

*you can use the time line on page 186*

**G** 

**H** 

**I** 

**J** 

**K** 

**L** 

**M** 

**N** 

# Temperatures 3

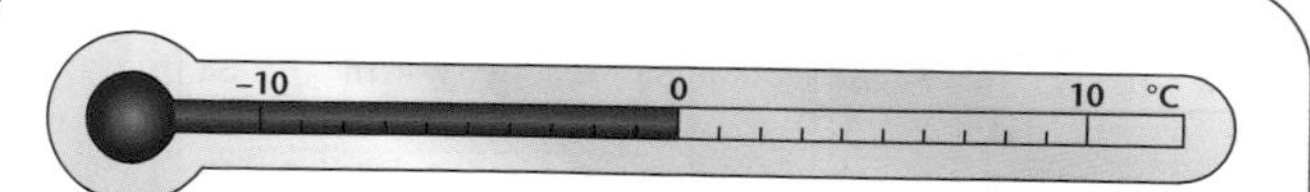


This thermometer shows **0 °C**.

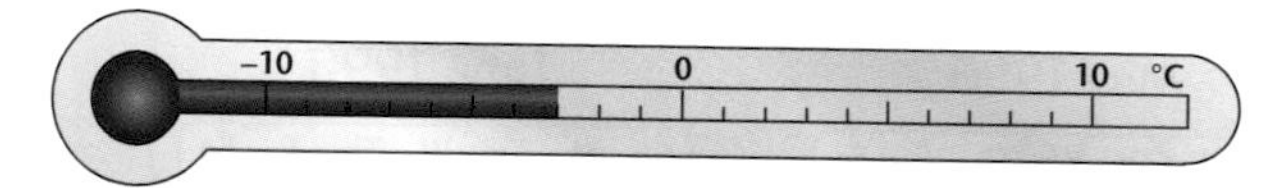


This thermometer shows
3 degrees less than zero.

We say –3 °C or minus three °C.

3 three
2 two
1 one
0 zero
–1 minus one
–2 minus two
–3 minus three
–4 minus four
–5 minus five
–6 minus six
–7 minus seven
–8 minus eight
–9 minus nine
–10 minus ten

▶ Write down the temperatures shown on these thermometers.

**1**

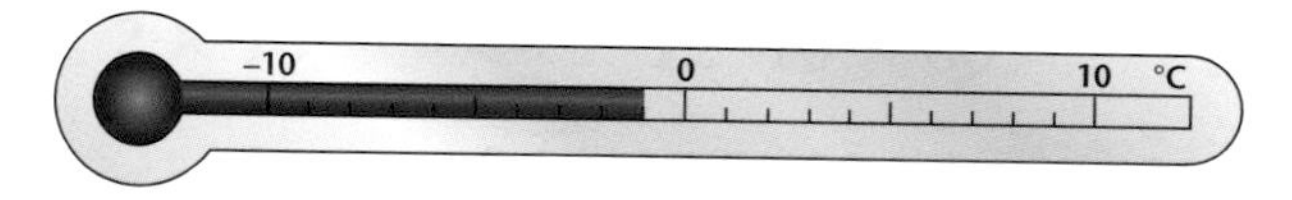


**2**

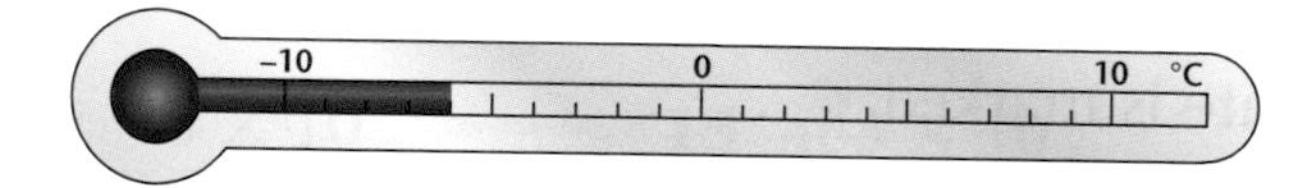


**3**

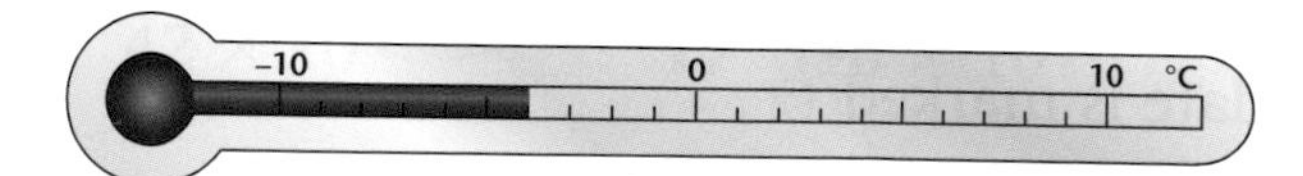


**4**

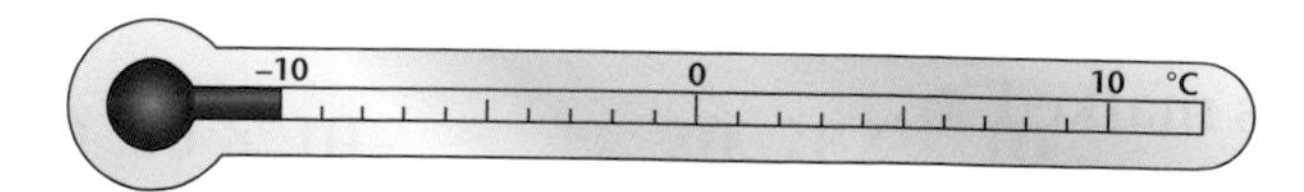

▶ Now answer these questions.

**5** **(a)** What is temperature A?
**(b)** What is temperature B?
**(c)** Which is colder, A or B?

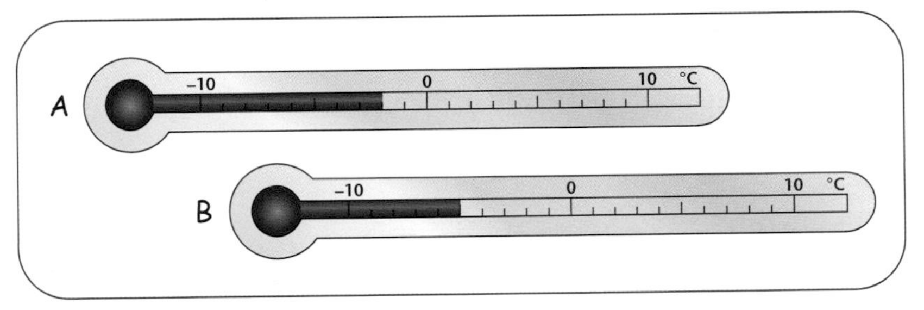


**6** **(a)** What is temperature L?
**(b)** What is temperature M?
**(c)** Which is warmer, L or M?

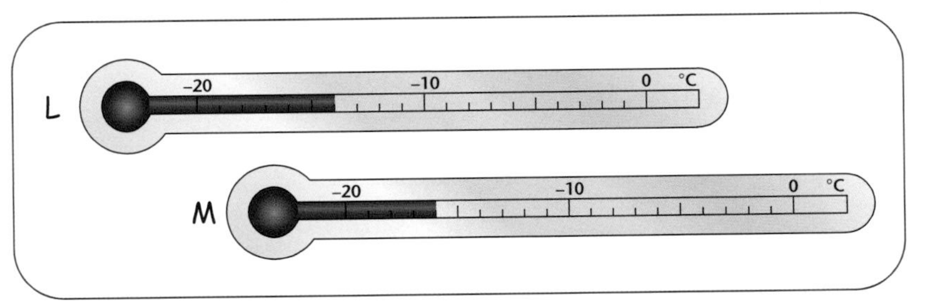


**7** Which is colder, P or Q?

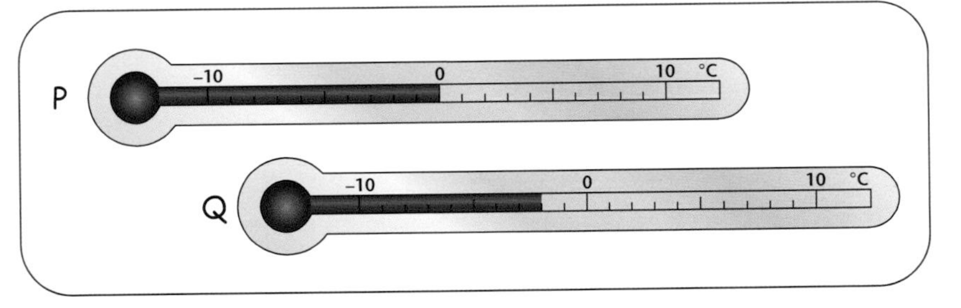


**8** Which is warmer, X or Y?

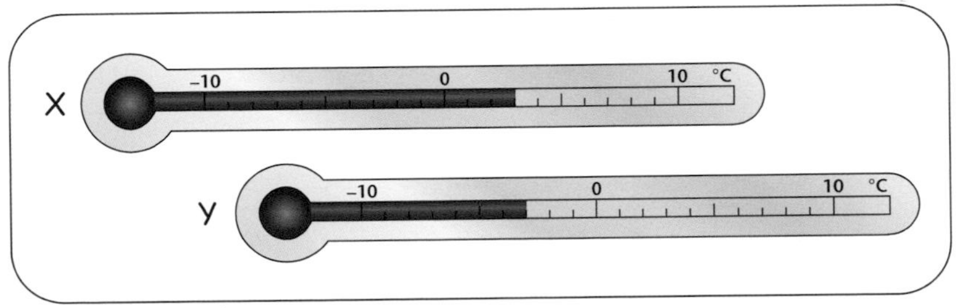

**9** This map shows the temperatures in some British cities.

**(a)** Which city is warmest?

**(b)** Which city is coldest?

**(c)** Which places are colder than 0 °C?

**(d)** Which places are warmer than 0 °C?

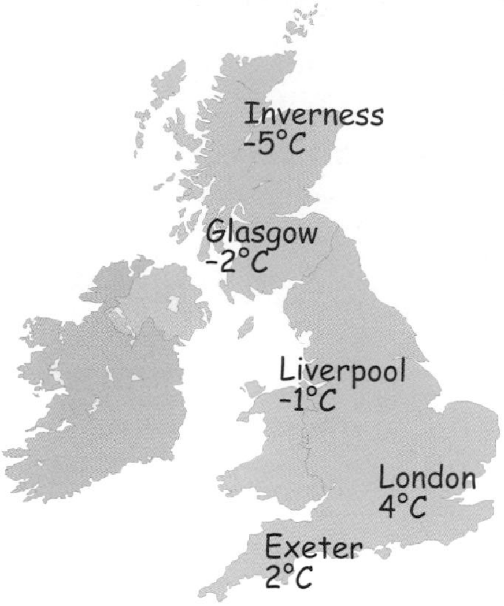

**10** This table shows the temperatures in some ski resorts.

| Klosters | −8 °C |
|---|---|
| Whistler | −15 °C |
| Aviemore | 1 °C |
| Chamonix | −5 °C |

**(a)** Which resort is the warmest?

**(b)** Which resort is the coldest?

**(c)** Write the resorts in order of temperature, starting with the coldest.
Use the number line to help you.

## Challenge

Find a weather report in a newspaper.

Are there any places where the temperature is below 0 °C?

What is the hottest place?
What is the coldest place?

Can you find the places on a map?

# The day trip

A community group organise a day trip to the seaside.

▶ Answer these questions.

**1** On the trip there are 30 pensioners, 15 school pupils and 10 adults.
How many people go on the trip altogether?

**2** They hire a coach that seats 64 passengers.
How many empty seats are there?

**3** The coach costs £110. They share the cost equally.
How much should each person pay for the coach?

**4** This is a map of the coach journey there and back.

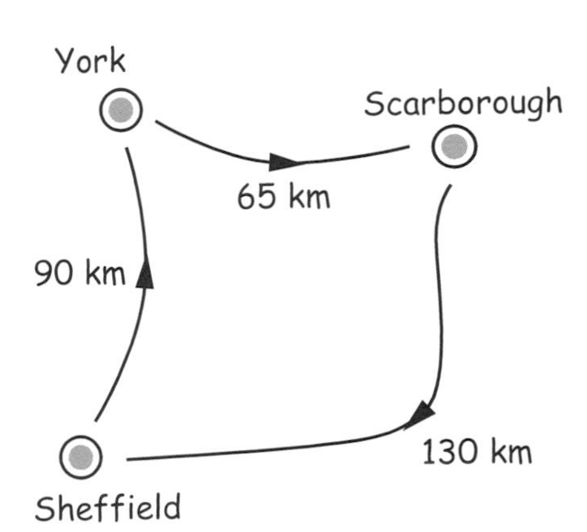


| | | |
|---|---|---|
| from | Sheffield | 08:30 |
| to | York | 09:30 |
| | ***shopping*** | |
| from | York | 12:00 |
| to | Scarborough | 13:00 |
| | ***lunch and beach*** | |
| from | Scarborough | 20:00 |
| to | Sheffield | 21:30 |

**(a)** On the way there the coach travels from Sheffield to York then to Scarborough.
How many kilometres is this altogether?

**(b)** How many kilometres shorter is the journey back?

**5** They have lunch together.

**(a)** How much does lunch cost for the 30 pensioners?

**(b)** How much does lunch cost altogether?

| **Set lunch** | |
|---|---|
| pensioners | £3 |
| children | £2 |
| adults | £5 |

**6** Use the table to answer these questions.

**(a)** How many hours do they have in Scarborough?

**(b)** How many hours do they have on the coach?

# Dog rescue

Mrs Carter runs the dog rescue centre.

**1** She uses 1·7 kilograms of dog food a day.
Each kilogram costs £2·39.

**(a)** How many kilograms of dog food does she use a week?

**(b)** How much does the dog food cost a week?

**2** Bouncer eats 3·5 kilograms of dog food a week.
Eddie eats 0·91 kilograms of dog food a week.

**(a)** How much dog food does Bouncer eat each day?

**(b)** How much dog food does Eddie eat each day?

**3** Chaser needs to walk at least 8·5 kilometres every day.
How far should he walk a week?

**4** Suzy walks Toby and Sam every day.
They walk 37·8 kilometres each week.

**(a)** How far do they walk a day?

**(b)** How far do they walk altogether in a weekend?

**5** Each kennel is 1·2 metres wide.
There are 6 kennels.
How wide is this altogether?

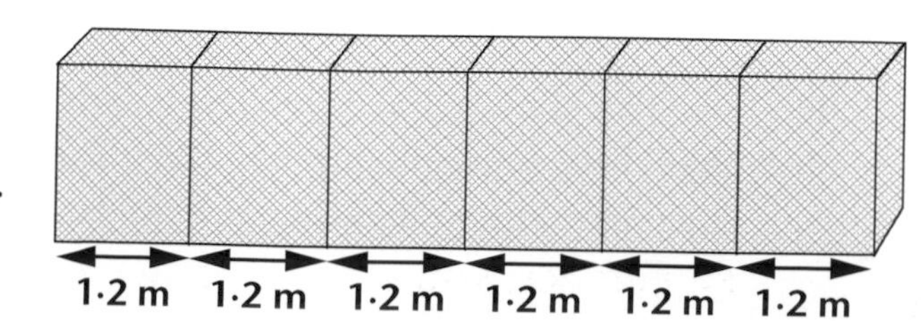

6 At Christmas a charity collects £67·44 to buy toys for the dogs.
How much money is this for each dog?

7 Mrs Carter has insurance to pay for visits to the vet.
She can pay £825·90 in advance for a year, or she can pay monthly or weekly.

*Pet insurance certificate*

Mrs Carter

| | |
|---|---|
| Annual fee: | £825·90 |
| Monthly fee: | £72·50 |
| Weekly fee: | £16·95 |

**(a)** How much does the insurance cost for a year if she pays 12 monthly payments?

**(b)** How much does the insurance cost for a year if she pays weekly payments?

## Target

*you need a calculator and a partner*

| **Step 1** | **Example** | **Scoring** |
|---|---|---|
| Your partner enters any two-digit number into the calculator. | 26 | less than 10 away score **1**<br>less than 5 away score **2** |
| **Step 2**<br>You multiply this by any number. | × 4.2 | less than 1 away score **5**<br>exactly 100 score **10** |
| The target is **100**. | 109.2 | |

► Take it in turns.
The first player to reach 20 points wins the game.
Remember: you only get one chance to hit the target.

# Unit 20

## Number patterns 4

### Review odd and even

▶ Copy the table.
Put these numbers in the right place in the table.

*Look back at page 78 for a reminder of odd and even.*

| odd | even |
|---|---|
| | |

26 15 37
42 9 34
8 51 0

### Ten times table

The numbers in the 10 times table all divide exactly by 10.

Look back at pages 102 and 198. All the numbers in the 10 times table end with a 0.

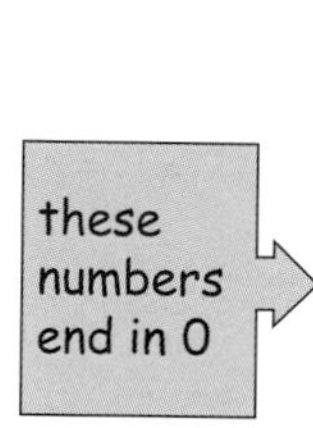


| divides exactly by 10 | does not divide by 10 |
|---|---|
| 30 470 160 90 600 | 38 841 252 74 |

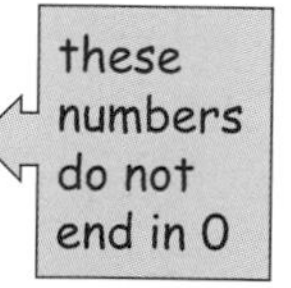


▶ Copy the table. Put these numbers in the right place in the table.

| divides exactly by 10 | does not divide by 10 |
|---|---|
| | |

68 20 43
30 9 100
792 35 805
29 300 70
10 471 60

### Five times table

The numbers in the 5 times table all divide exactly by 5.

Look back at page 83.
All the numbers in the 5 times table end with a 5 or a 0.

these numbers end in 5 or 0 →

| divides exactly by 5 | does not divide by 5 |
|---|---|
| 75 695 40 25 360 510 | 72 843 19 36 264 |

← these do not end in 5 or 0

▶ Copy the table. Put these numbers in the right place in the table.

| divides exactly by 5 | does not divide by 5 |
|---|---|
| | |

93 30 55
4 70 86
125 60 37
12 364 95
15 748 260

# Halves and quarters 3

▶ Work out these fractions with a calculator.

$\frac{1}{2}$ of 56:

**press 56 ÷ 2 =**

$\frac{1}{4}$ of 56:

**press 56 ÷ 4 =**

$\frac{3}{4}$ of 56:

**press 56 ÷ 4 x 3 =**

**A** $\frac{1}{2}$ of 150

**B** $\frac{1}{4}$ of 84

**C** half of 72

**D** $\frac{3}{4}$ of 200

**E** half of 64

**F** $\frac{1}{2}$ of 436

**G** $\frac{1}{4}$ of 720

**H** three quarters of 76

**I** $\frac{3}{4}$ of 508

**J** one quarter of 192

# Catalogues

This is part of a page from a mail-order catalogue.

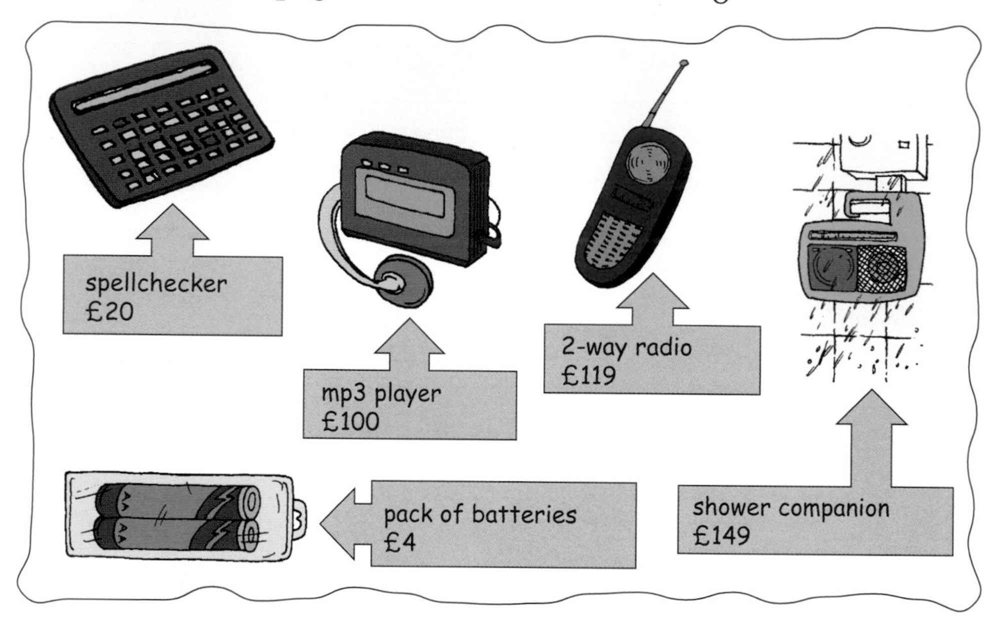


Eileen has ordered these items:

1 mp3 player
1 spellchecker
3 battery packs

This is how she starts to fill in the catalogue order form.

| Item | Quantity | Unit Price | Cost |
|---|---|---|---|
| mp3 player | 1 | £100 | £100 |
| spellchecker | 1 | £20 | £20 |
| | | | |
| | | | |
| | | Total cost: £ | |

 You need to practise filling in catalogue order forms . . . or use worksheet 20/1.

▶ This is a page from a charity catalogue.

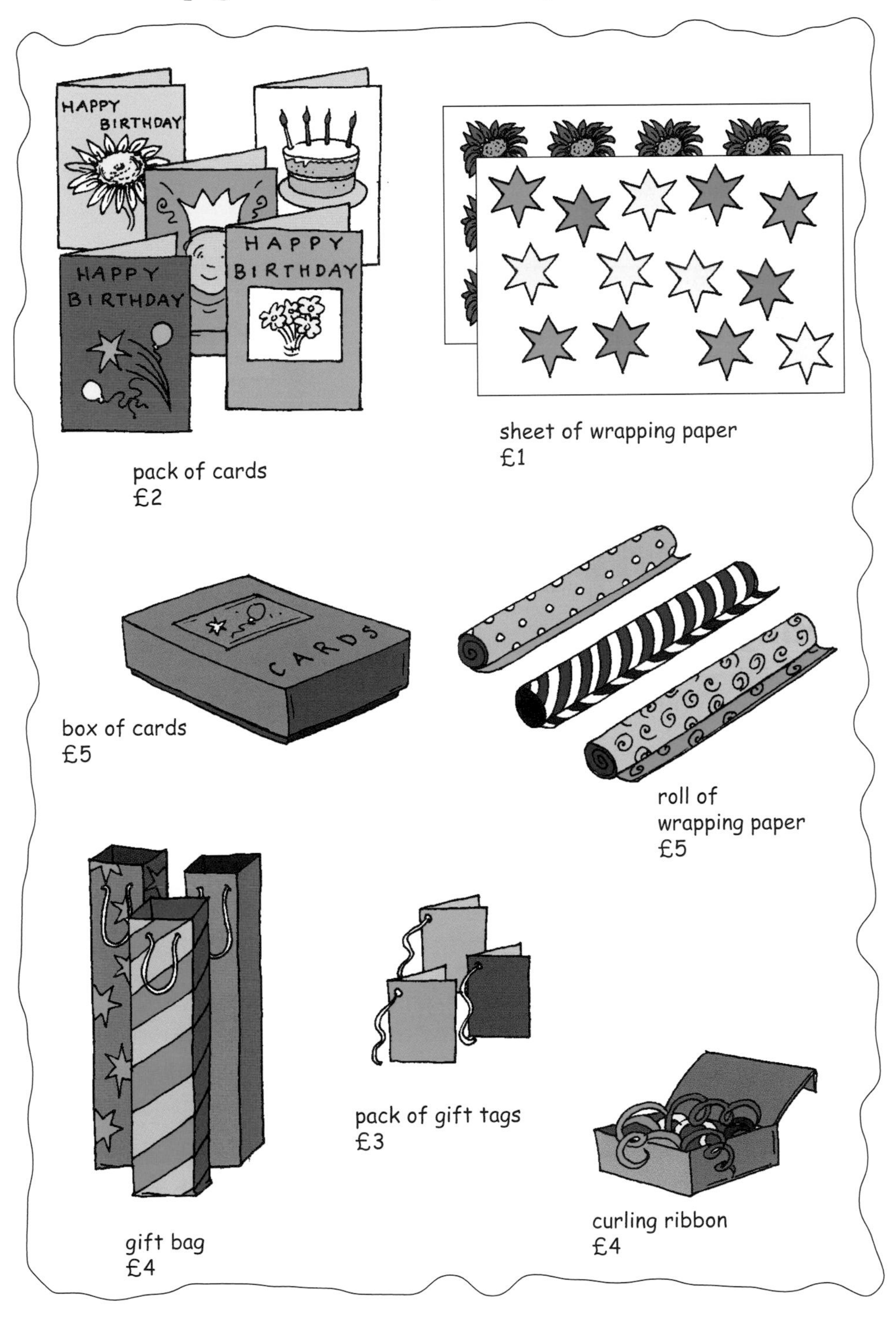


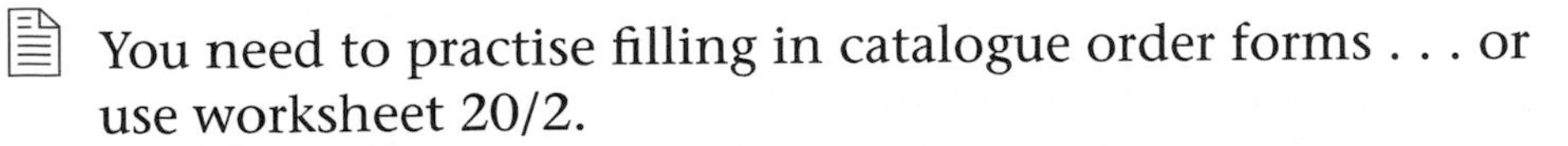
You need to practise filling in catalogue order forms . . . or use worksheet 20/2.

# Using units

centimetres grams minutes hours
kilometres kilograms seconds
millimetres pounds litres
metres pence millilitres

Units are used to measure different things.
You need to know which units are used for measuring each thing.

▶ Copy and complete these sentences.

We measure **money** in ____ (p) and ____ (£).

We measure **weight** in ____ (g) and ____ (kg).

We measure **liquid** in ____ (l) and ____ (ml).

We measure **time** in ____ (sec), ____ (min) and ____ (hr).

We measure **length** in ____ (m), ____ (mm), ____ (cm) and ____ (km).

▶ Which is the best unit to measure each of these?

**A** time to drive from London to Glasgow

**B** cost of a bottle of milk

**C** weight of a leopard

**D** height of a castle

**E** width of a ring

**F** amount of water to fill a watering can

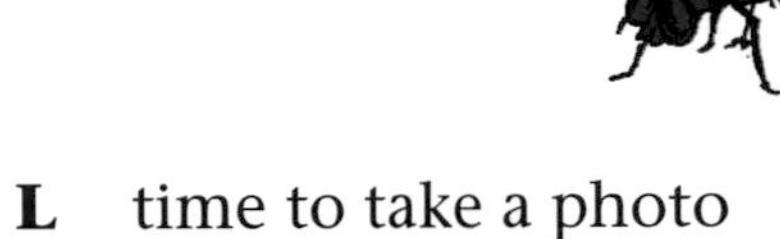

**G** cost of a pair of trainers

**H** time to cycle one kilometre

**I** distance from England to France

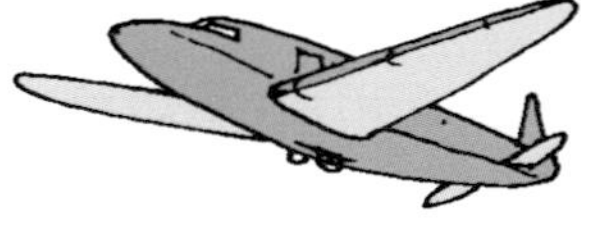

**J** height of a tree

**K** weight of a fly

**L** time to take a photo

**M** length of a table tennis bat

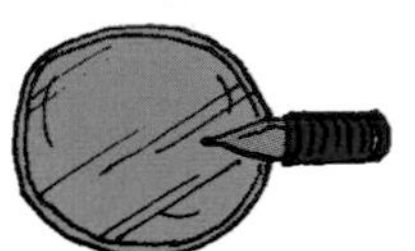

**N** amount of medicine on a spoon

**O** time to make a hot drink

**P** weight of a watch

# Rounding up and down

Muffins come in boxes of four.

Susan needs 25 muffins.

She works out 25 ÷ 4 to find out how many boxes to buy.

25 ÷ 4 = 6·25    6·25 is between 6 and 7

If she buys 6 boxes she will have 24 muffins.
This is not enough.

If she buys 7 boxes she will have 28 muffins.
She will have 3 left over.

Susan needs to buy 7 boxes of muffins.
She has 3 muffins extra.

Jack has 39 sweets to share between 5 people.

He works out 39 ÷ 5 to find out how many to give to each person.

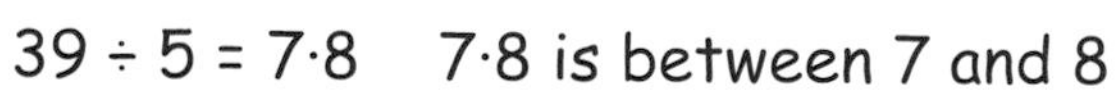
39 ÷ 5 = 7·8    7·8 is between 7 and 8

If he gives each person 7 sweets he will need 35 sweets.
He will have 4 left over.

If he gives each person 8 sweets he will need 40 sweets.
He has not got enough.

Jack gives each person 7 sweets. He has 4 sweets left over.

▶ Work these out. Think carefully about your answers.

You will need to round some answers **up** and some answers **down**.

**1** Sausages are sold in packs of eight. Mel needs 45 sausages.

**(a)** Work out 45 ÷ 8.

**(b)** Which two whole numbers does your answer come between?

**(c)** How many packs of sausages should Mel buy?

**(d)** How many extra sausages does she have?

**2** 

A group of 15 people are going by taxi to the station.
Four people can go in each taxi.

**(a)** How many taxis do they need?

**(b)** How many spare seats are there?

**3** Vera is packing toy cars. She puts 7 cars in each pack.
She has 50 toy cars to pack altogether.

**(a)** How many packs can she fill?

**(b)** How many spare cars does she have?

**4** Lee is planting bulbs in 4 tubs. He has 150 bulbs.
He wants to plant the same number of bulbs in each tub.

**(a)** How many bulbs can he plant in each tub?

**(b)** How many bulbs does he have left over?

**(c)** How many bulbs can he plant in each tub if he only uses 3 tubs?

**5** 

A group of 32 people goes into a cafe.
6 people can sit at each table.

**(a)** How many tables do they need?

**(b)** How many empty seats will there be?

**6** Emma is writing 45 Christmas cards.

**(a)** There are 8 cards in a pack.
How many packs does Emma need to buy?

**(b)** There are 10 stamps in a book.
How many books of stamps does Emma need to buy?

**7** At a record fair, CDs cost £6 each. Maisie has £50 to spend.

**(a)** How many CDs can she buy?

**(b)** How much money does she have left?

**8** Maisie has CD racks that hold 10 CDs each.
She now has 138 CDs altogether.
How many CD racks will she need?

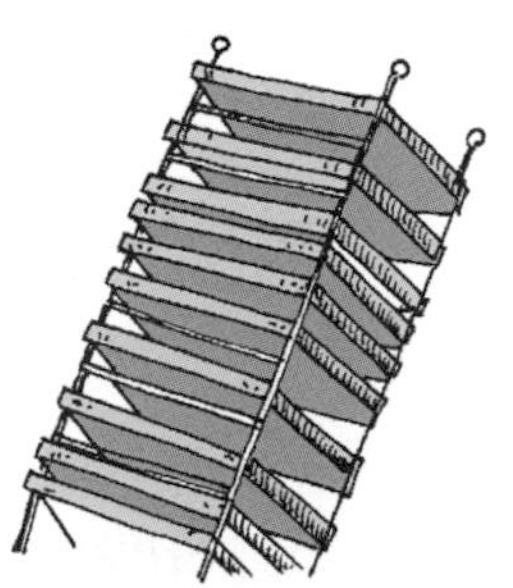

# Using diagrams

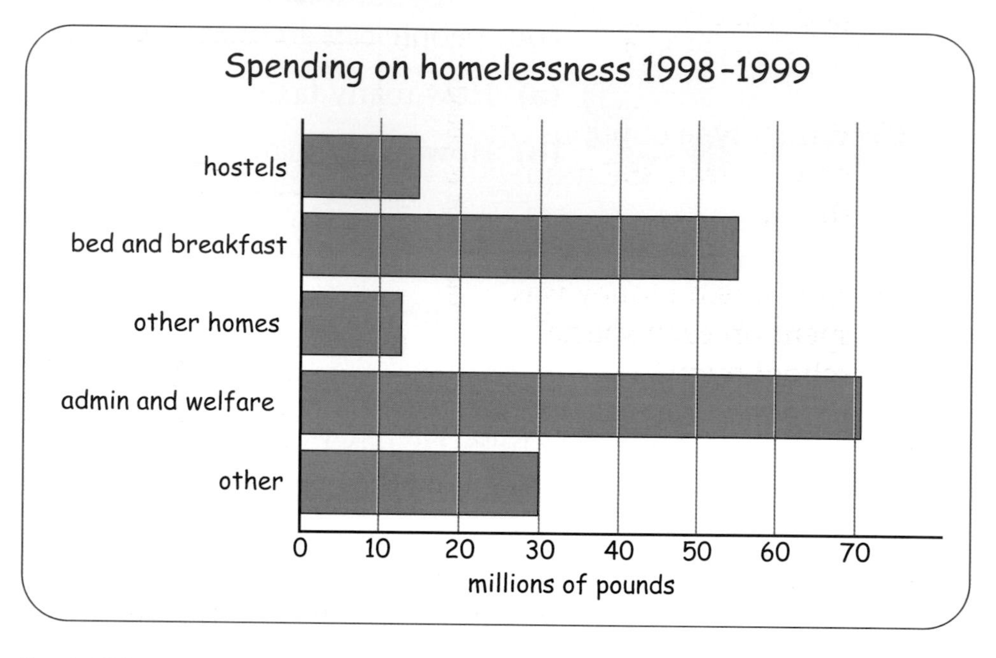


In 1998 to 1999, 14 million pounds was spent on hostels for the homeless.

**1** **(a)** What was £54 million spent on?

**(b)** What was the **most** money spent on?

**(c)** How much was spent on bed and breakfast?

The government spent 62 thousand million pounds on education in 1998 to 1999.

This pie chart shows how the money was spent.

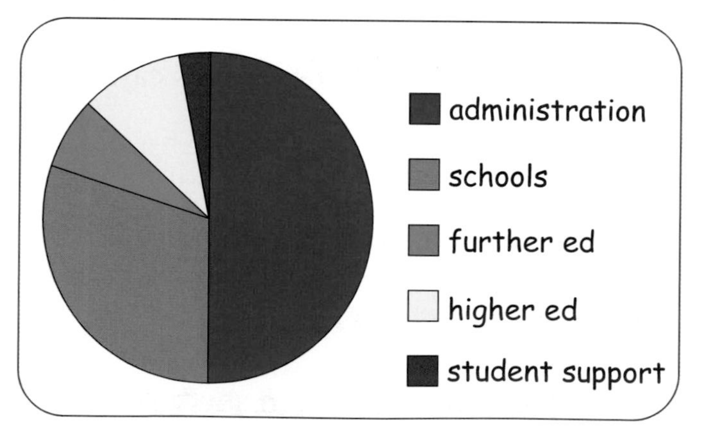


**2** **(a)** What was about half the money spent on?

**(b)** What was the least money spent on?

**(c)** Is **more than a quarter** or **less than a quarter** spent on schools?

**3** **(a)** Which type of pupil had just over £2000 spent on them?

**(b)** Which type of pupil had the least spent on them?

**(c)** How much money was spent on each special school pupil?

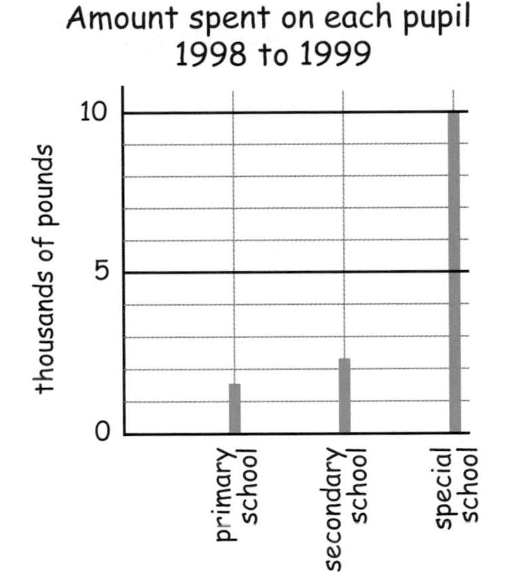


The environmental health department carry out inspections and serve a summons to anyone not obeying the law.

This pictogram shows the number of times a summons was served in England during 1998 to 1999.

| | Number served |
|---|---|
| Animal welfare | ▲▲▲▲▲▲▲▲▲▲▲▲▲ |
| Entertainment | ▲▲▲ |
| Food safety | ▲▲▲▲▲▲▲▲▲ |
| Health | ▲▲▲▲▲▲▲▲▲ |
| Housing | ▲▲▲▲▲ |
| Noise pollution | ▲▲▲▲▲▲▲ |

**Key:**
▲ stands for 100 served

**4** **(a)** How many were served for housing?

**(b)** How many were served for noise pollution?

**(c)** There were 2500 served to taxi or private hire firms. How many ▲ would be needed to show 2500?

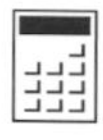

# Getting paid

per hour means for 1 hour

**1** John gets paid £6·25 per hour.
He works 8 hours.
How much does he get paid altogether for 8 hours?

**2** Tina gets paid £42·50 for 5 hours.
How much is this per hour?

**3** Ranjit was paid £6·57 an hour.
She gets a pay rise to £7·27 an hour.
How much extra is this?

**4** Rob worked at the seaside last summer.
This table shows how much Rob was paid each month.
How much was he paid altogether?

| Month | Pay |
|---|---|
| June | £648·52 |
| July | £827·39 |
| August | £952·60 |

**5**

Pat is paid £9·25 an hour.
She works 40 hours a week.
How much is she paid a week?

**6** Tim is paid £12·40 for each essay he types up.
On Monday he types up 14 essays.
How much is he paid for Monday?

**7** Dave is paid £327·95 for a 35 hour week.
How much is this per hour?

**8** Mr Jones pays £4·50 for each bag of empty aluminium cans he gets.
Anya sells him 4 bags.
Dan sells him 8 bags.
Jane sells him 13 bags.
Work out how much Mr Jones pays each person.

**9** Mary is paid £8·50 a week for her paper round.
She is saving the money to buy a stereo. It costs £127·50.
How many weeks will it take to save up £127·50?

**10** Alice, Chris and Rob earn money cleaning cars.
They charge £4·50 for each car.
On Saturday they cleaned 23 cars.

**(a)** How much money did they make altogether?

**(b)** They have to pay £23 for the cost of materials.
The rest is profit.
How much profit did they make?

**(c)** Alice did half the work so she gets half the profit.
How much does Alice get?

**(d)** Rob gets £15. Chris gets the rest of the profit.
How much does Chris get?

## Challenge

Look at the jobs section of the local papers.
Find a job that you would like to do.

How much is the pay?
How much would you get paid for a week . . . for a month . . . for a year . . . ?

# Reference

## Word lists

| | |
|---|---|
| 0 | zero |
| 1 | one |
| 2 | two |
| 3 | three |
| 4 | four |
| 5 | five |
| 6 | six |
| 7 | seven |
| 8 | eight |
| 9 | nine |
| 10 | ten |
| 11 | eleven |
| 12 | twelve |
| 13 | thirteen |
| 14 | fourteen |
| 15 | fifteen |
| 16 | sixteen |
| 17 | seventeen |
| 18 | eighteen |
| 19 | nineteen |
| 20 | twenty |
| 30 | thirty |
| 40 | forty |
| 50 | fifty |
| 60 | sixty |
| 70 | seventy |
| 80 | eighty |
| 90 | ninety |
| 100 | hundred |
| 1 000 | thousand |
| 1 000 000 | million |

number
digit
count

$+$ add
$+$ plus
carry

$-$ subtract
$-$ take away
$-$ minus
$-$ difference
borrow

$\times$ multiply
$\times$ times

$\div$ divide
$\div$ share
remainder

$=$ equals
total
answer
calculate
how many
altogether

problem
sum
question

once
twice
double

decimal
point

fraction
$\frac{1}{2}$ half
$\frac{1}{4}$ quarter

positive
negative

first
second
third
fourth
fifth

odd
even

multiple
divisible

pattern
repeat

**time**
second
minute
hour
day
week
fortnight
month
year
calendar

**money**
p pence
£ pounds

**length**
km kilometre
m metre
cm centimetre
mm millimetre

**weight**
kg kilogram
g gram

**liquid**
l litre
ml millilitre

**temperature**
° degrees
C Celsius

angle
shape
solid

circle
triangle
square
rectangle
pentagon
hexagon
octagon

parallel
straight
flat
curved
rounded
corners
pointed

cube
cuboid
cylinder
pyramid
sphere
cone

face
vertex

symmetry
reflection

the same as
more than
less than
greater
fewer

largest
larger
smallest
smaller
longer
longest
shorter
shortest
heavier
heaviest
lighter
lightest

north
south
east
west

clockwise
anticlockwise

left
right

above
below
behind
in front

tally
table
timetable
diagram
sorting

bar chart
pictogram
stick graph
pie chart

axis
axes
scale
frequency

most
least

## The hundred grid

| 1 | 2 | 3 | 4 | 5 | 6 | 7 | 8 | 9 | 10 |
|---|---|---|---|---|---|---|---|---|---|
| 11 | 12 | 13 | 14 | 15 | 16 | 17 | 18 | 19 | 20 |
| 21 | 22 | 23 | 24 | 25 | 26 | 27 | 28 | 29 | 30 |
| 31 | 32 | 33 | 34 | 35 | 36 | 37 | 38 | 39 | 40 |
| 41 | 42 | 43 | 44 | 45 | 46 | 47 | 48 | 49 | 50 |
| 51 | 52 | 53 | 54 | 55 | 56 | 57 | 58 | 59 | 60 |
| 61 | 62 | 63 | 64 | 65 | 66 | 67 | 68 | 69 | 70 |
| 71 | 72 | 73 | 74 | 75 | 76 | 77 | 78 | 79 | 80 |
| 81 | 82 | 83 | 84 | 85 | 86 | 87 | 88 | 89 | 90 |
| 91 | 92 | 93 | 94 | 95 | 96 | 97 | 98 | 99 | 100 |

## The tables grid

| 1 | 2 | 3 | 4 | 5 | 6 | 7 | 8 | 9 | 10 |
|---|---|---|---|---|---|---|---|---|---|
| 2 | 4 | 6 | 8 | 10 | 12 | 14 | 16 | 18 | 20 |
| 3 | 6 | 9 | 12 | 15 | 18 | 21 | 24 | 27 | 30 |
| 4 | 8 | 12 | 16 | 20 | 24 | 28 | 32 | 36 | 40 |
| 5 | 10 | 15 | 20 | 25 | 30 | 35 | 40 | 45 | 50 |
| 10 | 20 | 30 | 40 | 50 | 60 | 70 | 80 | 90 | 100 |